Non-Linear Programming

Mathematical Engineering, Manufacturing, and Management Sciences

Series Editor: Mangey Ram, Professor, Assistant Dean (International Affairs), Department of Mathematics, Graphic Era University, Dehradun, India

The aim of this new book series is to publish the rese arch studies and articles that bring up the latest development and research applied to mathematics and its applications in the manufacturing and management sciences areas. Mathematical tool and techniques are the strength of engineering sciences. They form the common foundation of all novel disciplines as engineering evolves and develops. The series will include a comprehensive range of applied mathematics and its application in engineering areas such as optimization techniques, mathematical modelling and simulation, stochastic processes and systems engineering, safety-critical system performance, system safety, system security, high assurance software architecture and design, mathematical modelling in environmental safety sciences, finite element methods, differential equations, reliability engineering, etc.

Circular Economy for the Management of Operations
Edited by Anil Kumar, Jose Arturo Garza-Reyes, and Syed Abdul Rehman Khan

Partial Differential Equations: An Introduction
Nita H. Shah and Mrudul Y. Jani

Linear Transformation
Examples and Solutions
Nita H. Shah and Urmila B. Chaudhari

Matrix and Determinant
Fundamentals and Applications
Nita H. Shah and Foram A. Thakkar

Non-Linear Programming
A Basic Introduction
Nita H. Shah and Poonam Prakash Mishra

For more information about this series, please visit: www.routledge.com/ Mathematical-Engineering-Manufacturing-and-Management-Sciences/book-series/ CRCMEMMS

Non-Linear Programming
A Basic Introduction

Nita H. Shah and Poonam Prakash Mishra

CRC Press
Taylor & Francis Group
Boca Raton London New York

CRC Press is an imprint of the
Taylor & Francis Group, an **informa** business

First edition published 2021
by CRC Press
6000 Broken Sound Parkway NW, Suite 300
Boca Raton, FL 33487-2742

and by CRC Press
2 Park Square, Milton Park, Abingdon, Oxon OX14 4RN

© 2021 Nita H. Shah and Poonam Prakash Mishra
CRC Press is an imprint of Taylor & Francis Group, LLC

Library of Congress Cataloging-in-Publication Data
Names: Shah, Nita H., author. | Mishra, Poonam Prakash, author.
Title: Non-linear programming: a basic introduction /
Nita H. Shah and Poonam Prakash Mishra. Description: First edition. |
Boca Raton, FL: CRC Press, an imprint of Taylor & Francis Group, LLC, 2021. |
Series: Mathematical engineering, manufacturing, and management sciences |
Includes bibliographical references and index.
Identifiers: LCCN 2020040934 (print) | LCCN 2020040935 (ebook) |
ISBN 9780367613280 (hardback) | ISBN 9781003105213 (ebook)
Subjects: LCSH: Nonlinear programming.
Classification: LCC T57.8 .S53 2021 (print) |
LCC T57.8 (ebook) | DDC 519.7/6–dc23
LC record available at https://lccn.loc.gov/2020040934
LC ebook record available at https://lccn.loc.gov/2020040935

ISBN: 978-0-367-61328-0 (hbk)
ISBN: 978-1-003-10521-3 (ebk)

Contents

Preface

Optimization is just an act of utilizing the given resources in best possible way. We use this concept knowingly or unknowingly in all aspects of life. Therefore, the term "optimization" has its wide range of applications in almost all the fields such as basic sciences, engineering and technology, business management, medical science and defence, etc. An optimization algorithm is a procedure which is executed iteratively by comparing various solutions till an optimum or a satisfactory solution is found. With the advent of computers, optimization has become a part of computer-aided activity. There are two types of algorithms widely used today as deterministic and stochastic algorithms.

In order to understand and explore this concept we need to see mathematical formulation behind it. Mathematically, an optimization problem consists of a function, better known as objective function that needs to be optimized (maximized/ minimized). There are some decision variables (design variables) on which value of objective function depends. Problem can be with or without constraints. If objective function as well as all the constraints are linear then that particular problem come under the category of linear programming problem (LPP) otherwise non-linear programming problem (NLP). If objective function or any of the constraint happens to be non-linear that problem is called NLP. Focus of this book is on NLP only. It is obvious that solving non-linear problems is more difficult compared to LPP. Choice of methods for solving an NLP depends on many parameters such as number of decision variables, concavity of the function, presence of constraint, equality or inequality constraints, and lastly overall complexity of the objective function in terms of continuity, smoothness, and differentiability. This book proposes a well synchronized and auto-guided approach for beginners to understand and solve different types of NLPs. This book has presented algorithms with their basic idea and appropriate illustrations for better understanding of the readers. Language and approach is simple to cater needs of undergraduate, postgraduate, and even research scholar in formulation and solution of their research problem. We have also mentioned MATLAB® syntax to use inbuilt functions of the MATLAB for solving different NLPs.

In this book we will discuss only non-linear programming (NLP). There are many conventional methods available in the literature for the optimization but still unable to approach all kinds of problem. So, researchers are continuously involved in developing new methods better known as optimization algorithms. Chapters of the book are as follows:

Chapter 1 discusses NLP for unimodal functions with single variable without constraints. Here we have discussed both conventional gradient-based methods and search algorithms for unimodal functions. These approaches act as fundamental for multivariable problems. We have also compared various approaches available to obtain solution and illustrate them briefly. Chapter 2 takes reader to the next level of multivariable NLP problems but without constraints. Here also we will demonstrate different approaches to solve the set of problems. This chapter also includes limitations of different methods which reader need to take care while applying them.

Chapter 3 allows reader to understand the most complex problems with non-linearity and multivariability in the presence of constraints. Here different conventional methods for equality and inequality constraints are explained. As this is most complex form, most of the time real-world problems cannot be addressed with conventional approaches. Therefore, some of the widely accepted modern approaches of stochastic search algorithms are mentioned in this chapter.

Chapter 4 makes reader understand the applicability of the above discussed methods in the different areas of pure sciences, engineering and technology, management, finance, etc. This part includes formulation of real-world problems into mathematical form that can be solved by any of the appropriate methods that allow readers to use these concepts in their research work widely. It is possible to compute program for all the algorithms mentioned in Chapters 1, 2, and 3 using C language or MATLAB. But, MATLAB also comes with built-in functions that can be simply called using appropriate syntax to approach most of the NLP. These inbuilt functions are discussed with their syntax for the convenience of readers.

Acknowledgement

First and foremost, I would like to thank **Almighty** for giving me all the strength and knowledge to undertake and complete the writing of this book successfully. I would like to thank **Prof. Nita H. Shah** from bottom of my heart for being my mentor and guide, since my Ph.D. tenure. She has always been a light house for my career and research activities. She has played the role of a friend, philosopher, and guide in my life. During the writing of this book she has helped me as a mentor as well as a co-author.

I am very grateful towards management of PDPU, SoT – Director – **Prof. S. Khanna** and Academic Director – **Prof. T. P. Singh** for giving me full support and mental space to write this book successfully. I would also like to extend my gratitude towards my departmental colleagues for supporting me consistently.

I sincerely express my gratitude to my parents **Mrs. Kanchan L. Pandey** and late **G. N. Pandey** for their generous blessings. At last, I would like to extend my special thanks to my true strength, my husband **Mr. Prakash Mishra** and my beloved daughter **Aarushi Mishra,** for giving me all love and affection to cherish my goals in life.

Author/Editor Biographies

Prof. Nita H. Shah received her Ph.D. in Statistics from Gujarat University in 1994. From February 1990 till now Prof. Nita is HOD of Department of Mathematics in Gujarat University, India. She is postdoctoral visiting research fellow of University of New Brunswick, Canada. Prof. Nita's research interests include inventory modeling in supply chain, robotic modeling, mathematical modeling of infectious diseases, image processing, dynamical systems and its applications, etc. Prof. Nita has published 13 monographs, 5 textbooks, and 475+ peer-reviewed research papers. Four edited books are prepared for IGI-global and Springer with coeditor as Dr. Mandeep Mittal. Her papers are published in high impact Elsevier, Inderscience, and Taylor and Francis journals. She is author of 14 books. By the Google scholar, the total number of citations is over 3070 and the maximum number of citation for a single paper is over 174. The H-index is 24 up to March 2020 and i-10 index is 74. She has guided 28 Ph.D. students and 15 M.Phil. students till now. Seven students are pursuing research for their Ph. D. degree. She has travelled to USA, Singapore, Canada, South Africa, Malaysia, and Indonesia for giving talks. She is Vice-President of Operational Research Society of India. She is council member of Indian Mathematical Society.

Dr. Poonam Prakash Mishra has completed her Ph.D in the year 2010 in mathematics. She also holds master's degree in business administration with specialization in operations management. Her core research area is modelling and formulation of inventory and supply chain management. She is also interested in the mathematical modelling of real-world problems with stochastic optimization. She has applied concepts of modelling and optimization in various fields such as for crude oil exploration, for sea ice route optimization problems, and impact of wind power forecasts on revenue insufficiency issue of electricity markets other than supply chain problems. She has successfully guided 03 students to earn their Ph.D. degree. She has more than 40 journal publications and 8 book chapters in various reputed international journals. She has successfully completed a funded project form SAC – ISRO and working on the other proposals.

Presently, she is working on Remote Sensing Investigation of Parameters that Affect Glacial Lake Outburst Flood (GLOF). She is working as a faculty of Mathematics at School of Technology – "Pandit Deendyal Petroleum University" at present.

1 One-Dimensional Optimization Problem

1.1 INTRODUCTION

In this section, we discuss the different methods available to optimize (minimize/maximize) the given function with only one variable. Methods used for one-dimensional optimization are highly useful for multivariable optimization. Firstly, we see the methods that can be used for unimodal functions. Unimodal functions are those functions that has only one peak or valley (in the given domain). Mathematically, function $f(x)$ is unimodal if (i) $x_1 < x_2 < x^*$ implies that $f(x_2) < f(x_1)$ and (ii) $x_2 > x_1 > x^*$ implies that $f(x_1) < f(x_2)$, where x^* is the minimum point. Figure 1.1 is a mind map that can help to explore available methods for these set of problems.

1.2 ANALYTICAL APPROACH

Analytic or conventional approach for extreme values can be achieved by following necessary and sufficient conditions, but this approach can work only for well-defined continuous and differentiable functions in the given domain.

Necessary condition: For a point x_0 to be the local extrema (local maximum and minimum) of a function $y = f(x)$ defined in the interval $a \leq x \leq b$ is that the first derivative of $f(x)$ exists as a finite number at $x = x_0$ and $f'(x_0) = 0$

Sufficient condition: If at an extreme point $x = x_0$ of $f(x)$, the first $(n-1)$ derivatives of it becomes zero, then:

(i) Local maximum of $f(x)$ occurs at $x = x_0$ if $f^{(n)}(x_0) < 0$, for n even
(ii) Local minimum of $f(x)$ occurs at $x = x_0$ if $f^{(n)}(x_0) > 0$, for n even
(iii) Point of inflection occurs at $x = x_0$ if, $f^{(n)}(x_0) \neq 0$, for n odd.

Algorithm:

1. Compute first derivative.
2. Solve the equation for x, $f'(x) = 0$, say $x = x_0$.

| OPTIMIZATION (MIN/MAX) FOR ONE-DIMENSIONAL UNIMODAL FUNCTIONS |

ANALYTICAL APPROACH (CALCULUS)

SEARCH ALGORITHMS

WITHOUT USING DERIVATIVE

- Unrestricted search
- Exhaustive search
- Dichotomous search
- Fibonacci search
- Golden section search
- Interpolation-based approach
  - Quadratic interpolation
  - Cubic interpolation

USING DERIVATIVE

- Newton method
- Secant method

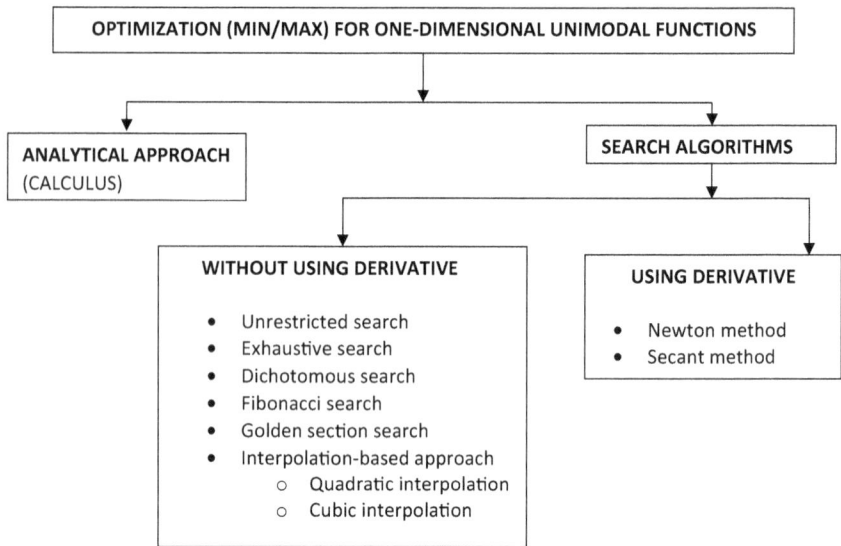

FIGURE 1.1   Tree diagram of available methods for one-variable optimization.

3. Compute $f''(x_0)$.
4. Declare x_0 as local minima if $f''(x_0) > 0$ or declare x_0 as local maxima if $f''(x_0) < 0$

Example:
If the total revenue (R) and total cost (C) function of a firm are given by $R = 30x - x^2$. and $C = 20 + 4x$, where x is the output. What is the maximum profit?

Solution
Let profit function be P, $P = R - C$
$P(x) = (30x - x^2) - (20 + 4x) = -x^2 + 26x - 20$
$P'(x) = -2x + 26 = 0 \Rightarrow x = 13$, using necessary condition,
$P''(x) = -2 < 0$, using sufficient condition $P(x)$ gives its maximum value at $x = 13$.
Hence, maximum profit is Rs. 149.

1.3   SEARCH TECHNIQUES

We shall discuss five different search techniques that are actually based on the assumption that function is unimodal, at least in the given range. This initial range is the interval of uncertainty and need to be made finer and finer with the iterations. These methods are also known as bracketing methods. These search techniques either simultaneously or step by step calculate the functional value at different points in the interval of uncertainty to obtain a finer interval of uncertainty in which minima/maxima lies.

Let $f(x)$ be our unimodal objective function and we desire to find minimum of this function in the range (x_F, x_L). Let x_F and x_L denote the first and last points of the interval of uncertainty. Then our initial interval of uncertainty is $L_0 = x_F - x_L$. Except unrestricted and exhaustive search techniques in all other search techniques, we shall try to reduce this initial level of uncertainty from $L_0, L_1, L_2 .... L_n$ in nth iterations. Let us see basic concepts, algorithm, and worked illustrations of each of the methods.

1.3.1 Unrestricted Search Technique

In this method functional value is calculated at an initial point say, x_0 then the following points can be calculated using fixed step length. This method can also be practiced with accelerated step size. Here we are presenting algorithm with fixed step length.

Algorithm:

1. Start with an initial guess point, say, x_0 in the given interval (x_F, x_L)
2. Find $f(x_0) = f_0$
3. Assume a step size s, find $x_1 = x_0 + s$
4. Find $f(x_1) = f_1$
5. If $f_1 < f_0$, and if the problem is for minimization, then unimodality of $f(x)$ indicates that the desired minimum cannot lie for $x > x_0$. Hence, the search can be continued further along points $x_2, x_3, x_4 .....$ using the assumption of unimodality while testing each pair of experiments. This procedure is continued until a point $x_i = x_1 + (i - 1)s$ shows an increase in the function value.
6. The search is terminated at x_i if $f_i > f_{i-1}$. In that case either x_i or x_{i-1} can be taken as the optimum point. Even midpoint of (x_i, x_{i-1}) can be considered as optimum point.

Example:
Find the minimum of $f = x(x - 2.5)$ by initial point as 1.01 and step size as 0.1 using unrestricted search algorithm.

Solution:
Let us calculate functional value at $x = 2$ and further taking step length as 0.2, as per the algorithm.

	Value	Functional value	Condition	Next iteration required
1	$x_0 = 1.01$	$f_0 = -1.5049$	-----	
2	$x_1 = 1.11$	$f_1 = -1.5429$	$f_0 > f_1$	Yes
3	$x_2 = 1.21$	$f_2 = -1.5609$	$f_1 > f_2$	Yes
4	$x_2 = 1.31$	$f_3 = -1.5589$	$f_2 < f_3$	No

This process shows that minimum lies between 1.21 and 1.31. Either of the point can also be chosen as minima. This algorithm can be repeated assuming smaller step size in the interval (1.21, 1.31).

1.3.2 EXHAUSTIVE SEARCH TECHNIQUE

This method evaluates the objective function at a predetermined number of equally spaced points in the interval (x_F, x_L). Let need to be evaluated at "n" equally spaced points in originally known interval of uncertainty of length $L_0 = x_L - x_F$ that divides L_0 into $n + 1$ intervals and we choose one of the interval as final interval of uncertainty called L_n. Let, say optimum value of function lies in the x_i th interval form the

n equally spaced intervals then $L_n = x_{j+1} - x_{j+1} = \dfrac{2}{n+1} L_0$

Algorithm:

1. Divide the given interval (x_F, x_L) into $(n + 1)$ equal intervals assuming "n" points within the given interval.
2. Find functional value simultaneously on all the $(n + 2)$ points.
3. Find a point x_j such that $f(x_{j-1}) > f\left(x_j\right) < f\left(x_{j+1}\right)$.
4. Declare x_j as minima.

Example:
Find the minimum of $f = x(x - 2.5)$ in the interval $(1, 1.4)$ using exhaustive search technique.

i	1	2	3	4	5	6	7	8	9
x_i	1	1.05	1.10	1.15	1.20	1.25	1.30	1.35	1.40
$f(x_i)$	-1.5000	-1.5225	-1.5400	-1.5525	-1.5600	-1.5625	-15600	-1.5525	-1.5400

Since $x_5 = x_7$, minimum lies between these values. Middle of these values, that is, $x_6 = 1.25$ can be considered as appropriate approximation.

1.3.3 DICHOTOMOUS SEARCH TECHNIQUE

In this method, initial interval of uncertainty L_0 is reduced in successive iteration by finding functional value at just two points. On the basis of functional value, interval of uncertainty is reduced.

Algorithm:

1. Let initial interval of uncertainty be $L_0 = (x_F, x_L)$. Assume very small value of delta δ.
2. Find two points x_1 and x_2 such that $x_1 = \dfrac{L_0}{2} - \dfrac{\delta}{2}$, $x_2 = \dfrac{L_0}{2} + \dfrac{\delta}{2}$.

3. If $f(x_2) < f(x_1)$, then we can discard (x_1, x_L). And our new interval of uncertainty is (x_F, x_1).
4. Continue Step 2 for the interval (x_F, x_1).
5. Let the new points be x_3 and x_4.
6. If $f(x_3) < f(x_4)$ then the new interval of uncertainty be (x_F, x_4).
7. Continue the process to reduce the interval of uncertainty to the desirable level.

Example:
Find the minimum for $f(x) = x(x - 2.5)$ in the interval $(1, 1.4)$ using dichotomous search.

Here, $f(x) = x(x - 2.5)$, let us find x_1 and x_2 using the above-mentioned formula by taking $\delta = 0.001$

Iteration-1

$$x_1 = \frac{L_0}{2} - \frac{\delta}{2} = \frac{2.4}{2} - 0.0005 = 1.1995$$

$$x_2 = \frac{L_0}{2} + \frac{\delta}{2} = \frac{2.4}{2} + 0.0005 = 1.2005$$

Now, $f_1 = -1.5599$ and $f_2 = -1.5600$. Since $f_1 > f_2$ interval (x_F, x_1) can be discarded. So, next interval of uncertainty is $(x_1, x_L) = (1.1995, 1.4)$.
Let us find x_3 and x_4 using the same method.

Iteration-2

$$x_3 = \frac{L_1}{2} - \frac{\delta}{2} = \frac{2.5995}{2} - 0.0005 = 1.29925$$

$$x_4 = \frac{L_1}{2} + \frac{\delta}{2} = \frac{2.3395}{2} + 0.0005 = 1.30025$$

Now, $f_3 = -1.5600, f_4 = -1.5599$. Since $f_3 < f_4$, interval $(x_4, x_L) = (1.30025, 1.4)$ can be discarded. So, next interval of uncertainty is $(x_1, x_L) = (1.1995, 1.30025)$
Let us find x_5 and x_6.

Iteration-3

$$x_5 = \frac{L_2}{2} - \frac{\delta}{2} = \frac{2.49975}{2} - 0.0005 = 1.249375$$

$$x_6 = \frac{L_2}{2} + \frac{\delta}{2} = \frac{2.49975}{2} + 0.0005 = 1.259375$$

Now, $f_5 = -1.562499$, $f_6 = -1.562412$. Since $f_5 < f_6$, interval $(x_6, x_4) = (1.259375, 1.30025)$ can be discarded. So, next interval of uncertainty is $(x_6, x_4) = (1.1995, 1.259375)$.

So, minimum lies between the interval (1.1995, 1.259375) at the end of three iterations. The midpoint 1.22943 can be considered as the required optimal value.

Interval of uncertainty reduces in this way after every iteration

$$L_1 = \frac{1}{2}(L_0 + \delta), \quad L_2 = \frac{1}{2}\left(\frac{1}{2}(L_0 + \delta)\right) + \frac{\delta}{2}, \quad L_3 = \left(\frac{1}{2}\left(\frac{1}{2}(L_0 + \delta)\right) + \frac{\delta}{2}\right) + \frac{\delta}{2} \cdots\cdots$$

$$\cdots\cdots\cdots L_n = \frac{L_n}{2^{n/2}} + \delta\left(1 - \frac{1}{2^{n/2}}\right)$$

1.3.4 Fibonacci Search Method

This is also an iterative method that uses the sequence of Fibonacci numbers, $\{Fn\}$, for placing the experiments. These numbers are defined as $F_0 = F_1 = 1, F_n = F_{n-1} = F_{n-2}$, $n = 2, 3, 4, \ldots$

which gives 1, 1, 2, 3, 5, 8, 13, 21, 34, 55, 89 … ..

Other than function with initial interval of uncertainty, this algorithm needs value of n to execute the algorithm.

Algorithm:

1. Let initial interval of uncertainty is $L_o = [a, b]$, and number of experiments be n.
2. Calculate x_1 and x_2 between [a, b] such that using x_1 and x_2 is L^* units apart from a and b respectively, $x_1 = a + L^*, \quad x_1 = b + L^*$ where $L^* = \frac{F_{n-2}}{F_n} L_0$.
3. Calculate f_1 and f_2. If $f_1 < f_2$, then we shall discard (x_2, b) and form $L_o = [a, b]$ and new interval of uncertainty will be $L_2 = [a, x_2]$. If $f_1 > f_2$ then (a, x_1) will get discarded and new interval of uncertainty will be $[x_1, b]$.
4. Calculate x_3 using $x_3 = a + (x_2 - x_1)$ considering $L_2 = [a, x_2]$ as present interval of uncertainty, such that distance between x_3 and a is same as x_2 and x_1.
5. Calculate f_1 and f_3, if $f_1 > f_3$, then discard (a, x_1) and next interval of uncertainty will be $[x_1, x_2]$.
6. Calculate x_4 using $x_4 = x_1 + (x_2 - x_3)$.
7. Repeat Step 5 till x_n.

Example:
Find the minimum for $f(x) = x(x - 5.3)$ in the interval [1, 4] using Fibonacci search algorithm.

Solution
Here function is $f(x) = x(x - 5.3)$, $L_0 = [1, 4]$. Let us assume $n = 6$

Let us calculate x_1 and x_2 using $x_1 = a + L^*, \quad x_1 = b + L^*$ where $L^* = \frac{F_{n-2}}{F_n} L_0$

$$\left. \begin{array}{l} x_1 = a + L^* = 1 + 1.153846 = 2.153846 \\ x_2 = b - L^* = 4 - 1.153846 = 2.846154 \end{array} \right\} \quad \text{where} \quad L^* = \frac{5}{13}(3) = 1.153846$$

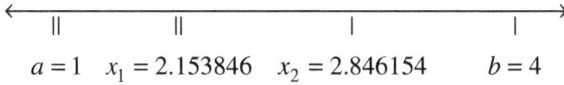

$$
\begin{array}{cccc}
\| & \| & | & | \\
a = 1 & x_1 = 2.153846 & x_2 = 2.846154 & b = 4
\end{array}
$$

$f_1 = -67759738, f_2 = -6.9840236$. Since $f_1 > f_2$, we will discard (a, x_1) and next interval of uncertainty is $L_1 = [x_1, b] = [2.153846, 4]$.

Let us calculate x_3 using $x_3 = x_1 + (b - x_2) = 2.153846 + (4 - 2.846154) = 3.307692$.

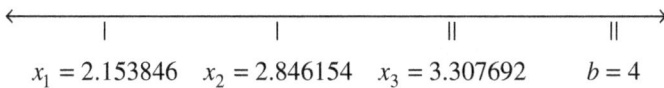

$$
\begin{array}{cccc}
| & | & \| & \| \\
x_1 = 2.153846 & x_2 = 2.846154 & x_3 = 3.307692 & b = 4
\end{array}
$$

Here, $f_2 = -6.9840236, f_3 = -6.5899412$. Since, $f_2 < f_3$, we will discard (x_3, b) and next interval of uncertainty is $L_2 = [x_1, x_3]$.

Let us calculate x_4 using $x_4 = x_1 + (x_3 - x_2) = 2.153846 + (3.307692 - 2.846154) = 2.615384$

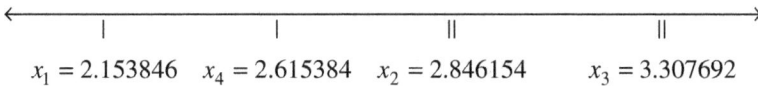

$$
\begin{array}{cccc}
| & | & \| & \| \\
x_1 = 2.153846 & x_4 = 2.615384 & x_2 = 2.846154 & x_3 = 3.307692
\end{array}
$$

Here, $f_4 = -7.0213017$. Since $f_2 > f_4$, we will discard (x_2, x_3) and new interval of uncertainty is $L_3 = [x_1, x_2]$.

Let us calculate x_5 using $x_5 = x_1 + (x_2 - x_4) = 2.153846 + (2.846154 - 2.615384) = 2.384616$

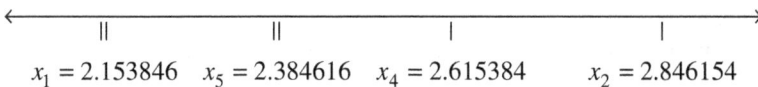

$$
\begin{array}{cccc}
\| & \| & | & | \\
x_1 = 2.153846 & x_5 = 2.384616 & x_4 = 2.615384 & x_2 = 2.846154
\end{array}
$$

Here, $f_5 = -6.9520713$. Since $f_5 > f_4$, we will discard (x_1, x_5) and new interval of uncertainty is $L_3 = [x_5, x_2]$.

Let us calculate x_6 using $x_6 = x_1 + (x_4 - x_5) = 2.153846 + (2.615384 - 2.384616) = 2.384614$

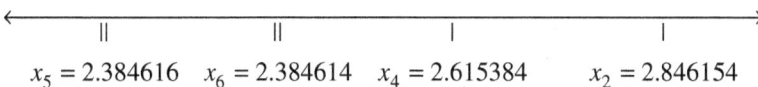

$$
\begin{array}{cccc}
\| & \| & | & | \\
x_5 = 2.384616 & x_6 = 2.384614 & x_4 = 2.615384 & x_2 = 2.846154
\end{array}
$$

Here, $f_6 = -6.9520703$. Since $f_6 > f_4$ we will discard (x_5, x_6) and new interval of uncertainty is $[x_6, x_2] = 0.23077 = L_6$

Here, $L_6 / L_0 = 0.23077/3 = 0.0769233$ (Reduction ratio for $n = 6$).

1.3.5 GOLDEN SECTION SEARCH METHOD

This method is again an iterative method that allows deletion of some part of an interval of uncertainty in every iteration on the basis of unimodality and functional value of the function. But, it does not need a predefined "n" number of iterations as in the case of Fibonacci. It allows to terminate the process as per the tolerance and considers very large "n". Therefore, simply applying condition "n tends to infinity" on Fibonacci can as well lead to this ratio. Golden ratio is being used since ancient times by Greek architects and engineers in designing process. They believed that any construction with side a and b that satisfies the ratio $\dfrac{a+b}{a} = \dfrac{a}{b} = \gamma$ carries happiness and prosperity.

$$1 + \frac{b}{a} = \frac{a}{b} = \gamma \Rightarrow 1 + \frac{1}{\gamma} = \gamma \Rightarrow \gamma^2 - \gamma - 1 = 0$$

Positive root of the equation is $\gamma = 1.61803$. This is the golden ratio and we will use this in the golden section search method.

Algorithm:

1. Given interval of uncertainty is $L_0 = [x_F, x_L]$.
2. Find $L_2^* = \dfrac{1}{\gamma^2} L_0$ and using this obtain $x_1 = x_F + L_2^*$ and $x_2 = x_L - L_2^*$.
3. Find f_1 and f_2. If $f_1 > f_2$ then discard $[x_F, x_1]$ and declare $[x_1, x_L]$ as new interval of uncertainty.
4. Next experiment is x_3, which needs to be obtained using the equation $x_3 = x_1 + (x_L - x_1)$.
5. Continue this process till desired tolerance is obtained.

Example:
Find the minimum for $f(x) = x(x - 5.3)$ in the interval $[1, 4]$ using golden section search algorithm.

Solution:

Initial level of tolerance is $L_0 = [1,4]$. Let us find $L_2^* = \dfrac{1}{\gamma^2} L_0 = 0.382(3) = 1.146$

Now, $x_1 = x_F + L_2^* = 1 + 1.146 = 2.146$ and $x_2 = x_L - L_2^* = 4 - 1.146 = 2.854$. $f_1 = -6.7684, f_2 = -6.9808$. Since $f_1 > f_2$ we can discard $[x_F, x_1] = [1, 2.146]$ and next interval of uncertainty will be $[x_1, x_L] = [2.146, 4]$.

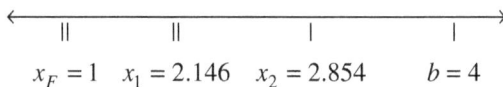

$$x_F = 1 \quad x_1 = 2.146 \quad x_2 = 2.854 \quad b = 4$$

TABLE 1.1
Comparison of various search techniques

Method	Formula	$n = 4$	$n = 10$
Exhaustive search	$L_n = \dfrac{2L_0}{n+1}$	$L_n = (0.4)L_0$	$L_n = (0.18182)L_0$
Dichotomous search	$L_n = \dfrac{L_0}{2^{n/2}} + \delta\left(1 - \dfrac{1}{2^{n/2}}\right),$ $\delta = 0.01$	$L_n = (0.25)L_0$ $+ 0.0075$	$L_n = (0.03125)L_0$ $+ 0.0096875$
Fibonacci search	$L_n = \dfrac{L_0}{F_n}$	$L_n = (0.2)L_0$	$L_n = (0.11245)L_0$
Golden section search	$L_n = (0.618)^{n-1}L_0$	$L_n = (0.236)L_0$	$L_n = (0.01315)L_0$

Now next experiment is $x_3 = x_1 + (x_L - x_2) = 2.146 + (4 - 2.854) = 3.292$.

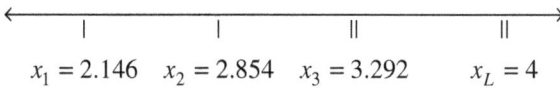

$$x_1 = 2.146 \quad x_2 = 2.854 \quad x_3 = 3.292 \qquad x_L = 4$$

Here, $f_3 = -6.610336$, $f_2 < f_3$. Hence, we will discard $[x_3, x_L] = [3.292, 4]$ and new interval of uncertainty be [2.146, 2.854]. Process can be continued to attain better approximation.

Table 1.1 shows a tabular comparison of various approaches of search techniques in order to understand the efficiency of different methods. Efficiency of search techniques depends on L_n / L_0. Here L_0 represent original interval of uncertainty whereas L_n represent reduce interval of uncertainty after n iterations. Comparison clearly shows that Fibonacci and Golden section searches reduce this ratio faster compared to exhaustive search and dichotomous search techniques. As soon as we move to higher iterations, Fibonacci proves to be a better search technique than Golden section.

1.3.6 Interpolation Method (Without Using Derivative)

In this section we will see how interpolation can be used to find minima of given unimodal function without actually finding its derivative. It comprises of two methods: (1) quadratic interpolation and (2) cubic interpolation methods.

1.3.6.1 Quadratic Interpolation

For a given unimodal function $f(x)$, $x \in [x_F, x_L]$ whose minima need to be obtained is approximated with a quadratic polynomial $p(x)$. In this approach $p(x)$ gets optimized in the interval $[x_F, x_L]$ instead of $f(x)$, $x \in [x_F, x_L]$. Optimum value x^* can be accepted if it satisfies $\left| \dfrac{f(x^*) - p(x^*)}{f(x^*)} \right| < \varepsilon$ where ε is a very small predefined value.

Algorithm:

1. Initialize with x_F, x_1, x_L such that $x_F < x_1 < x_L$ where x_1 can be midpoint of the given interval $[x_F, x_L]$.
2. Approximate given function $f(x)$ with quadratic polynomial $p(x) = a_0 + a_1 x + a_2 x^2$ using following set of equations.

$$f(x_F) = a_0 + a_1(x_F) + a_2(x_F)^2$$

$$f(x_1) = a_0 + a_1(x_1) + a_2(x_1)^2$$

$$f(x_L) = a_0 + a_1(x_L) + a_2(x_L)^2$$

$$a_0 = \frac{f(x_F)x_1 x_L(x_L - x_1) + f(x_1)x_L x_F(x_F - x_L) + f(x_L)x_1 x_F(x_1 - x_F)}{(x_F - x_1)(x_1 - x_L)(x_L - x_F)}$$

$$a_1 = \frac{f(x_F)(x_1^2 - x_L^2) + f(x_1)(x_L^2 - x_F^2) + f(x_L)(x_F^2 - x_1^2)}{(x_F - x_1)(x_1 - x_L)(x_L - x_F)}$$

$$a_2 = \frac{-f(x_F)(x_1 - x_L) + f(x_1)(x_L - x_F) + f(x_L)(x_F - x_1)}{(x_F - x_1)(x_1 - x_L)(x_L - x_F)}$$

Obtain optimal x using $x^* = \dfrac{-a_1}{2a_2}$

3. Check $\left| \dfrac{f(x^*) - p(x^*)}{f(x^*)} \right| < \varepsilon$

4. There will be four cases on the basis of value x^* and x_1 and their functional values. Obtain the next interval of uncertainty as per the cases:

CASE 1 $\Rightarrow x^* < x_1, f\left(x^*\right) < f(x_1)$, then new interval be $[x_F, x_1]$.

Declare new $x_F = x_F,\ x^* = x_1,\ x_L = x_1$

CASE 2 $\Rightarrow x^* < x_1, f\left(x^*\right) > f(x_1)$, then new interval be $[x^*, x_L]$.

Declare new $x_F = x^*,\ x_1 = x_1,\ x_L = x_1$

CASE 3 $\Rightarrow x^* > x_1, f\left(x^*\right) < f(x_1)$, then new interval be $[x_1, x_L]$.

Declare new $x_F = x_1,\ x_1 = x^*,\ x_L = x_L$

CASE 4 $\Rightarrow x^* > x_1, f\left(x^*\right) > f(x_1)$, then new interval be $[x_F, x^*]$.

Declare new $x_F = x_F,\ x_1 = x_1,\ x_L = x^*$

5. Continue to refine the interval of uncertainty till desired approximation is achieved.

Example:

Find the minimum value of the function $f(x) = 0.5 - x \tan^{-1}\left(\dfrac{1}{x}\right) - \dfrac{0.9}{1 + x^2}$ using quadratic interpolation method in the interval of $[0.45, 0.65]$.

Solution

Here $x_F = 0.45$, $x_1 = 0.55$, $x_L = 0.65$. Let $\varepsilon = 0.001$

Iteration-1:

Solve for $p(x) = a_0 + a_1 x + a_2 x^2$ using following set of equations

$$f(0.45) = a_0 + a_1(0.45) + a_2(0.45)^2$$
$$f(0.55) = a_0 + a_1(0.55) + a_2(0.55)^2$$
$$f(0.65) = a_0 + a_1(0.65) + a_2(0.65)^2$$

Value of $a_0 = -0.548247, a_1 = -0.76669, a_2 = -0.6333$

$$x^* = \frac{-a_1}{2a_2} = \frac{0.77669}{0.6333} = 0.245938$$

$$\left| \frac{f(x^*) - p(x^*)}{f(x^*)} \right| = \left| \frac{-0.675677 + 0.698499}{-0.675677} \right| = 0.03377 > \varepsilon$$

Now, $f_1 = -0.778353, f^* = -0.675677$. Since $f_1 < f^*$.
New interval will be $[x^*, x_L] = [0.2459, 0.65]$

Iteration 2:

$$x_F = 0.2459, x_1 = 0.55, x_L = 0.65$$

Solving for $p(x) = a_0 + a_1 x + a_2 x^2$ using following set of equations

$$f(0.2459) = a_0 + a_1(0.2459) + a_2(0.2459)^2$$
$$f(0.55) = a_0 + a_1(0.55) + a_2(0.55)^2$$
$$f(0.65) = a_0 + a_1(0.65) + a_2(0.65)^2$$

$$x^* = \frac{-a_1}{2a_2} = \frac{0.989916}{2(0.819324)} = 0.405530$$

$$\left| \frac{f(x^*) - p(x^*)}{f(x^*)} \right| = \left| \frac{-0.75366275 + 0.748452}{-0.75366275} \right| = 0.00691 > \varepsilon. \quad \text{Since} \quad f_1 < f^*. \quad \text{New}$$
interval will be $[x^*, x_L] = [0.405530, 0.65]$

Iteration 3:

$$x_F = 0.40553, x_1 = 0.55, x_L = 0.65$$

$$x^* = \frac{-a_1}{2a_2} = \frac{1.31452}{2(1.08983)} = 0.603084$$

$$\left|\frac{f(x^*) - p(x^*)}{f(x^*)}\right| = \left|\frac{-0.78 + 0.780758}{-0.78}\right| = 0.0009717 < \varepsilon$$

We have achieved the desired accuracy and can terminate the process. Optimal (minima) value for given function is 0.603084

1.3.6.2 Cubic Interpolation

Cubic interpolation approach is analogous to quadratic interpolation. In quadratic interpolation, we approximate the given unimodal function $f(x)$, $x \in [x_F, x_L]$ whose minima need to be determined by a cubic polynomial $p(x)$. Similarly, in cubic interpolation approach $p(x) = a_0 + a_1 x + a_2 x^2 + a_3 x^3$ gets optimized in the interval $[x_F, x_L]$ instead of $f(x)$, $x \in [x_F, x_L]$. Optimum value x^* can be accepted if it satisfies $\left|\frac{f(x^*) - p(x^*)}{f(x^*)}\right| < \varepsilon$ where ε is a very small predefined value.

1.4 GRADIENT-BASED APPROACH

Calculus suggests that necessary conditions for a given function $f(x)$ to have minimum at x^* is $f'(x^*) = 0$. In this section we shall find the root of equation $f'(x^*) = 0$ using Newton method and Secant method.

1.4.1 NEWTON METHOD

Newton's method, also known as the Newton–Raphson method, is a method of finding roots. Using this concept we can find root of $f'(x)$ and this point would be local minima of $f(x)$. In case initial approximation is not close to x^*, it may diverge.

Let quadratic approximation of the function $f(x)$ at $x = x^*$ using Taylor's be

$$f(x) = f\left(x^*\right) + (x - x^*)f'\left(x^*\right) + (x - x^*)^2 f''\left(x^*\right) + \ldots$$

Deriving once and equating to zero gives

$$f'(x) = f'\left(x^*\right) + (x - x^*)f''\left(x^*\right) = 0$$

Algorithm:

Step 0: Set x_0, (initial approximation), $\varepsilon > 0$, k = 0, 1, 2 ... .

Step 1: Get $x_{k+1} = x_k - \frac{f'(x_k)}{f''(x_k)}$

Step 2: If $|f'(x_k)| < \varepsilon$ declare x_k as optimal point or repeat Step 1.

Example:
Find minimum of the function $f(x) = x^4 - x^3 + 5$ using Newton method.
Take $x_0 = 1$

Solution:
We have $f(x) = x^4 - x^3 + 5$, then $f'(x) = 4x^3 - 3x^2$ and $f''(x) = 12x^2 - 6x$.

First iteration: $x_1 = x_0 - \dfrac{f'(x_0)}{f''(x_0)} = 1 - \dfrac{1}{6} = 0.833$

Second iteration: $x_1 = x_1 - \dfrac{f'(x_1)}{f''(x_1)} = 0.833 - \dfrac{0.23037}{3.32866} = 0.7637$

Minima for the given function is 0.7637

1.4.2 SECANT METHOD

This method is again used to approximate root of $f'(x)$ that eventually happens to be minima of $f(x)$. Here, we use secant to approximate the roots instead of tangent.

Let us have two points 'a' and 'b' on the function $f'(x)$ such that $f'(a).f'(b) < 0$. Then, we can evaluate the next approximation using secant $(a, f'(a))$ and $(b, f'(b))$

as $x_k = b - \left[\dfrac{b-a}{f'(b) - f'(a)} \right] f'(b)$.

For next iteration, if $f'(a).f'(x_k) < 0$, $(a, f'(a))$ and $(x_k, f'(x_k))$ will be new secant.

Otherwise, $f'(b).f'(x_k) < 0$ allows $(b, f'(b))$ and $(x_k, f'(x_k))$ to form new secant. Continue the process till the desired accuracy is achieved.

Algorithm:

Step 0: Set $[a, b]$ (initial approximation) such that $f'(a).f'(b) < 0$, $\varepsilon > 0$, $k = 0$, 1, 2 ... .

Step 1: Get $x_k = b - \left[\dfrac{b-a}{f'(b) - f'(a)} \right] f'(b)$

Step 2: If $f'(a).f'(x_k) < 0$ then $(a, f'(a))$, $(x_k, f'(x_k))$ is new secant, otherwise if $f'(b).f'(x_k) < 0$ then $((b, f'(b))$, $(x_k, f'(x_k))$ is new secant.

Step 3: If $|f'(x_k)| < \varepsilon$ declare x_k as optimal point or repeat Step 1.

TRY YOURSELF

Q1. Find the minimum of the function $f(x) = x^5 - 5x^3 - x + 25$ using the following methods:
   (a) Unrestricted search technique using initial interval of uncertainty as (0, 3)
   (b) Dichotomous search technique using initial interval of uncertainty as (0, 3). Take $\delta = 0.001$

(c) Golden section search technique using initial interval of uncertainty as (0, 3)

(d) Fibonacci search algorithm using initial interval of uncertainty as (0, 3)

Answer: 1.751

Q2. Find the minimum of the function $f(x) = \dfrac{x/2}{\log(x/3)}$ using the following methods taking initial guess as $x_0 = 8.5$.

(a) Quadratic interpolation method
(b) Cubic interpolation method
(c) Newton method
(d) Secant method

Answer: 8.155

Q3. Find the number of experiments to be conducted in the following methods to obtain a value of $\dfrac{1}{2}\left(\dfrac{L_n}{L_0}\right) = 0.01$:

(a) Exhaustive search
(b) Dichotomous search with $\delta = 0.001$
(c) Fibonacci method
(d) Golden section method

Answer (a) n ≥ 99, (b) n ≥ 14, (c) n ≥ 9, n ≥ 10

Q4. Find the maximum of the function $f(x) = \dfrac{6x}{x^2 - 3x + 5}$ using the following methods taking initial guess as $x_0 = 2.5$. Comment which method converges faster to the optimum.

(a) Quadratic interpolation method
(b) Cubic interpolation method
(c) Newton method
(d) Secant method

Answer: 2.236

2 Unconstrained Multivariable Optimization

2.1 INTRODUCTION

In this section, we will discuss several methods with their algorithm to solve unconstrained multivariable problems for optimization. We will also demonstrate working of these algorithms with suitable examples. There are two approaches to find optimal solution for multivariable problems. If function is smooth and can be derived gradient-based methods are followed, which are also known as indirect methods. However, if this is not the case then optimal point of the given function can be obtained by different search algorithms popularly known as direct search methods. Tree diagram in Figure 2.1 illustrates the different methods that are available for solving multivariable optimization problems.

2.2 DIRECT SEARCH METHODS

These methods are applicable to functions that are not differentiable and may not be continuous. Different search techniques using various logics approximate the extreme points. Eventhough there are many techniques available under direct search methods, we will discuss simplex method, Hooke–Jeeves method, and Powell's method in detail. We will also understand key points behind random search, grid search, and univariate search as these are basic and less commonly used methods.

$$\underset{x_1, x_2 \ldots\ldots\, x_n}{Minimize}\ f(x_1, x_2 \ldots\ldots x_n)$$

In general, search algorithms have the following structure:

1. $x^{K+1} = x^K + \Delta x^K d^K$, where d^K is direction and x^K is increment.
2. We need to start with an initial guess and must specify termination criterion.

```
┌─────────────────────────────────────────────────────────────┐
│   OPTIMIZATION (MIN/MAX) FOR MULTIVARIABLE PROBLEM          │
└─────────────────────────────────────────────────────────────┘
```

DIRECT SEARCH METHODS

- Random Search Method
- Grid Search Method
- Univariate Search Method
- Pattern Search Algorithm
  - i) Hooke-Jeeves Method
  - ii) Powell's Method
- Simplex Method

INDIRECT METHODS

(Gradient-based methods)

- Using Hessian Matrix
- Steepest Descent Method
- Newton's Method
- Quasi Method

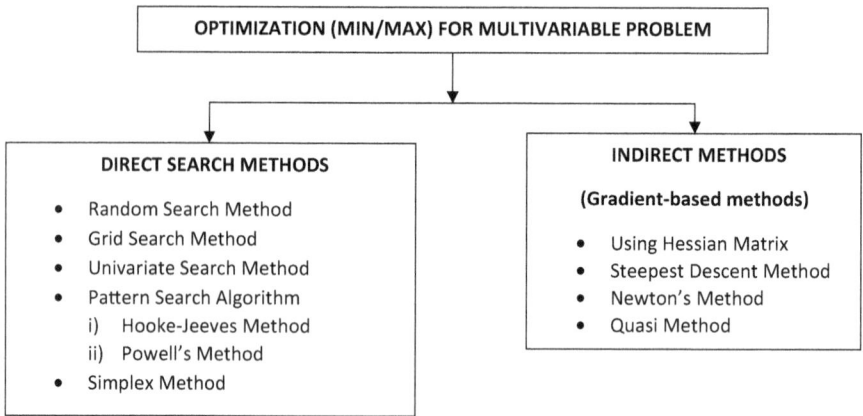

FIGURE 2.1 Tree diagram of available methods for multivariable optimization problems.

2.2.1 RANDOM SEARCH METHOD

As name suggests here functional value is calculated for the bounded decision variables. Let $x_i, i \in [1,n]$ are decision variables with lower and upper bounds as l_K and u_K respectively for the objective function $Minimize\ f(x_1, x_2 \ldots\ldots x_n)$.

$$x_1, x_2 \ldots\ldots x_n$$

Approach of random search uses random numbers $0 < \{r_{11}, r_{12}, r_{13} \ldots\ldots r_{1n}\} < 1$, to obtain first approximation by $X_1 = \begin{Bmatrix} x_1 \\ x_2 \\ \vdots \\ x_n \end{Bmatrix} = \begin{Bmatrix} l_1 + r_{11}(u_1 - l_1) \\ l_2 + r_{12}(u_2 - l_2) \\ \vdots \\ l_n + r_{1n}(u_n - l_n) \end{Bmatrix}$

Similarly, new set of random numbers can be generated $0 < \{r_{21}, r_{22}, r_{23} \ldots\ldots r_{2n}\} < 1$.

And X_2 can be obtained by $X_2 = \begin{Bmatrix} l_1 + r_{21}(u_1 - l_1) \\ l_2 + r_{22}(u_2 - l_2) \\ \vdots \\ l_n + r_{2n}(u_n - l_n) \end{Bmatrix}$

Now, obtain $f(X_1), f(X_2) \ldots f(X_n) \ldots$. Further get $(f(X_K) = Min\{f(X_1), f(X_2) \ldots f(X_n) \ldots\})$.

This method is also suitable for functions with discontinuous and non-differentiable points and can find both local and global extremes. Still this method is not efficient as it needs too many functional values to reach to conclusion but it can be fused with other methods to obtain the global optimum.

2.2.2 GRID SEARCH METHOD

This method divides the given space into grids and further value of objective function is calculated at each node to have an idea of extremes. An objective function with two variables will have a design grid in two-dimension and are easy to interpret. For example, a two-design variable space with four partitions have $2^4 = 16$ nodes but a design space with let say five design variables will have $5^4 = 625$ nodes. It is obvious that computation cost is too high in case of problems with high number of decision variables and this method too is not an efficient search method to find the optimal solution.

2.2.3 UNIVARIATE SEARCH METHOD

In this method, only one variable vary at one point of time keeping other variables constant. Value of one variable can be optimized by any of the methods discussed in Chapter 1.

Algorithm:

Step 0: Set x_0, $k = 0,1,2...$
Step 1: Find the search direction d_k as

$$d_k^T = \begin{cases} (1,0,0......0)\ for\ i=1,\ n+1,\ 2n+1...... \\ (0,1,0......0)\ for\ i=2,\ n+2,\ 2n+2...... \\ (0,0,1......0)\ for\ i=2,\ n+3,\ 2n+3...... \\ .... \\ (0,0,0......0)\ for\ i=1,\ 2n+3n...... \end{cases}$$

Step 2: Find $f_k = f(x_k), f^+ = f(x_k + \varepsilon d_k), f^- = f(x_k - \varepsilon d_k)$

If $\begin{cases} f^+ < f_k, \text{ then } d_k \text{ is correct direction,} \\ f^- < f_k, \text{ then } -d_k \text{ is correct direction} \end{cases}$ for a minimization problem

Step 3: Find optimal step length $\Delta x_k *$ such that $f(x_k \pm \Delta x_k^* d_k) = \min_{\Delta x_k}(x_k \pm \Delta x_k d_k)$.
Step 4: Set $x_{k+1} = x_k \pm \Delta x_k^* d_k$ and $f(x_{k+1}) = f_{k+1}$
Step 5: Set the value of $k = k+1$ and go to step 1. Continue the process till desired accuracy is achieved.

Example:
Find *Min* $f(x_1, x_2) = x_1 - x_2 + 2x_1^2 + 2x_1 x_2 + x_2^2$ using univariate method.

Solution:

Let us set $x_0 = \begin{bmatrix} 0 \\ 0 \end{bmatrix}$, $\varepsilon = 0.01$.

Step 1: let $d_0 = \begin{bmatrix} 1 \\ 0 \end{bmatrix}$

Step 2: $f_0 = 0$, $f^+ = f(x_0 + \varepsilon d_0) = f\left(\begin{bmatrix} 0 \\ 0 \end{bmatrix} + 0.01 \begin{bmatrix} 1 \\ 0 \end{bmatrix}\right) = f\begin{bmatrix} 0.01 \\ 0 \end{bmatrix} = 0.0102$

$$f^- = f(x_0 - \varepsilon d_0) = f\left(\begin{bmatrix} 0 \\ 0 \end{bmatrix} - 0.01 \begin{bmatrix} 1 \\ 0 \end{bmatrix}\right) = f\begin{bmatrix} -0.01 \\ 0 \end{bmatrix} = -0.0098$$

Since, $f^- < f_0, -d_0$ will be direction for minimization.

Step 3: $f(x_0 + \Delta x_0 d_0) = f\left(\begin{bmatrix} 0 \\ 0 \end{bmatrix} - \Delta x_0 \begin{bmatrix} 1 \\ 0 \end{bmatrix}\right) = f\begin{bmatrix} -\Delta x_0 \\ 0 \end{bmatrix} = 2(\Delta x_0)^2 - \Delta x_0.$

$$\frac{df}{d\Delta x_0} = 0 \Rightarrow \Delta x_0^* = 0.25$$

Thus $x_1 = x_0 - \Delta x_0^* d_0 = \begin{bmatrix} 0 \\ 0 \end{bmatrix} - 0.25 \begin{bmatrix} 1 \\ 0 \end{bmatrix} = \begin{bmatrix} -0.25 \\ 0 \end{bmatrix}$

Here, $f_1 = -0.125$

Step 4: Choose the search direction $d_1 = \begin{bmatrix} 0 \\ 1 \end{bmatrix}$

$$f^+ = f(x_1 + \varepsilon d_1) = f\left(\begin{bmatrix} -0.25 \\ 0 \end{bmatrix} + 0.01 \begin{bmatrix} 0 \\ 1 \end{bmatrix}\right) = f\begin{bmatrix} -0.25 \\ 0.01 \end{bmatrix} = -0.1399$$

$$f^- = f(x_1 - \varepsilon d_1) = f\left(\begin{bmatrix} -0.25 \\ 0 \end{bmatrix} - 0.01 \begin{bmatrix} 0 \\ 1 \end{bmatrix}\right) = f\begin{bmatrix} -0.25 \\ -0.01 \end{bmatrix} = -0.1099.$$

Since $f^+ < f_1, d_1$ is correct choice for the direction.

Step 5: $f(x_1 + \Delta x_1 d_1) = f\left(\begin{bmatrix} -0.25 \\ 0 \end{bmatrix} - \Delta x_1 \begin{bmatrix} 0 \\ 1 \end{bmatrix}\right) = f\begin{bmatrix} -0.25 \\ \Delta x_1 \end{bmatrix}$

$$= (\Delta x_1)^2 - (1.5)\Delta x_1 - 0.375.$$

$$\frac{df}{d\Delta x_1} = 0 \Rightarrow \Delta x_1^* = 0.75$$

Thus $x_2 = x_1 + \Delta x_1^* d_1 = \begin{bmatrix} -0.25 \\ 0 \end{bmatrix} + 0.75 \begin{bmatrix} 0 \\ 1 \end{bmatrix} = \begin{bmatrix} -0.25 \\ 0.75 \end{bmatrix}$

$f_2 = -0.6875$

2.2.4   PATTERN SEARCH ALGORITHM

There are certain methods that use pattern directions and are called Pattern Search Algorithm. In this section we shall be discussing Hooke–Jeeves method and Powell's method. This algorithm helps to search the minimum along the pattern direction S_i defined by $S_i = X_i - X_{i-1}$, where, X_i = point at the end of n univariate steps.

X_{i-1} = beginning of n univariate steps

2.2.4.1   Hooke–Jeeves Method

It is a sequential technique in which each step consists of two kinds of moves:

(a) Exploratory move – It is performed in the neighbourhood of current point using univariate technique to explore local behaviour of objective function. Aim of exploratory move is to obtain best possible point around the current point.
(b) Pattern move – It is performed along a pattern direction. Pattern direction can be achieved by considering the current best points as well as the previous points using the formula $x_p^{(k+1)} = x^{(k)} + \left(x^{(k)} - x^{(k-1)} \right)$, but if new point is not the improved point then we need to re-explore the exploratory move with small step length.

Algorithm:

Step 0: Define starting point $x^{(0)}$; increment $\Delta_i = 1, 2, \ldots\ldots n$; step reduction factor $\alpha > 1$; $\varepsilon > 0$ (termination factor). Set k=0.

Step 1: Perform exploratory search with $x^{(k)}$ as base point. Let x be output of exploratory move. If exploratory move is a success then $x^{(k)} = x$ and go to step 2.

Step 2: If $\|\Delta_i\| < \varepsilon$, stop; current solution $\approx x^*$. Else set $\Delta_i = \Delta_i / \alpha$ and go to step 1.

Step 3: Set $k = k + 1$ and perform pattern move: $x_p^{(k+1)} = x^{(k)} + \left(x^{(k)} - x^{(k-1)} \right)$.

Step 4: Perform exploratory search with $x_p^{(k+1)}$ as base point. Say, output $= x^{(k+1)}$.

Step 5: If $f(x^{(k+1)}) < f(x^{(k)})$, go to step 3 or else go to step 2.

Example:

Find the minimum of $f(x, y) = \left(x^2 + y - 11 \right)^2 + (x + y^2 - 7)^2$. Let the initial approximation be $X^{(0)} = (x^{(0)}, y^{(0)}) = (0,0)$

Solution:

1st Iteration

Let $X^{(0)} = (x^{(0)}, y^{(0)}) = (0,0)$, $\Delta_1 = 0.5$

Let us go for first exploratory move in x-direction

$$(0,0) \Rightarrow (0+0.5, 0) = (0.5, 0)$$
$$(0,0) \Rightarrow (0,0)$$
$$(0,0) \Rightarrow (0-0.5, 0) = (-0.5, 0)$$

Functional value at possible moves is

$$f(0.5, 0) = 157.8, f(0,0) = 170, f(-0.5, 0) = 171.81,$$

Since, $f(0.5, 0) = 157.8$, New $(x^{(0)}, y^{(0)}) = (0.5, 0)$ is least we will use this to move in y-direction.

$$(0.5,0) \Rightarrow (0.5,0+0.5) = (0.5,0.5)$$
$$(0.5,0) \Rightarrow (0.5,0)$$
$$(0.5,0) \Rightarrow (0.5,0-0.5) = (0.5,-0.5)$$

Functional value at possible moves is

$$f(0.5,0.5) = 144.12, f(0.5,0) = 157.81, f(0.5,-0.5) = 165.62,$$

Since, $f(0.5,0.5) = 144.12$, Finally, $X_E^{(1)} = (0.5,0.5)$
Now we will apply pattern move by

$$X_p^{(k+1)} = X^{(k)} + \left(X^{(k)} - X^{(k-1)}\right) = 2X^{(k)} - X^{(k-1)}$$
$$X_p^{(2)} = 2(0.5,0.5) - (0,0) = (1,1)$$

2nd Iteration: let $X^{(2)} = (1,1)$. We shall move on with x-direction

$$(1,1) \Rightarrow (1+0.5,1) = (1.5,1)$$
$$(1,1) \Rightarrow (1,1) = (1,1)$$
$$(1,1) \Rightarrow (1-0.5,1) = (0.5,1)$$

$$f(1.5,1) = 80.3125, f(1,1) = 106, f(0.5,1) = 125.3$$

We will choose $f(1.5,1) = 80.3125$ to explore in y-direction.

$$(1.5,1) \Rightarrow (1.5,1+0.5) = (1.5,1.5)$$
$$(1.5,1) \Rightarrow (1.5,1) = (1.5,1)$$
$$(1.5,1) \Rightarrow (1.5,1-0.5) = (1.5,0.5)$$

$$f(1.5,1.5) = 63.12, f(1.5,1) = 80.31, f(1.5,0.5) = 95.67$$

Now, $X_E^{(2)} = (1.5,1.5)$. Let us compute pattern move using last two moves

$$X_p^{(3)} = 2(1.5,1.5) - (0.5,0.5) = (2.5,2.5)$$

Continuing the process we will get $X_p^{(4)} = (4.5,2.5)$,
We have $f(X^{(1)}) = 144.12, f(X^{(2)}) = 144.12, f(X^{(3)}) = 0, f(X^{(4)}) = 50$.

Since, $f(X^{(3)}) < f(X^{(4)})$, we can terminate it and start with new $\Delta = \dfrac{\Delta}{2} = (0.25,0.25)$.

2.2.4.2 Powell's Method

This method is also known as Powell's Conjugate Direction Method and is one of the most widely used direct search method. It is a unique extension of basic pattern search method. Basically, this method minimizes a quadratic function in a finite number of steps with the fact that any non-linear function can be approximated with a quadratic function near its minimum.

Powell's method generates n linearly independent search directions from the previous best point, and then unidirectional search is performed along each of these search directions. In general, non-linear unconstrained multivariable function generates more than one go for n unidirectional searches; whereas, Powell's method finds minimum of a quadratic function by one go at a time of n unidirectional searches along each search direction. This works on conjugate directions instead of arbitrary search directions. Two directions $d^{(1)}$ and $d^{(2)}$ are called conjugate with respect to positive definite matrix Q if $\left(d^{(1)}\right)^T Q\left(d^{(2)}\right) = 0$.

Let Q be an $(n \times n)$ square positive definite symmetric matrix. A set of n linearly independent search directions $d^{(i)}$ is called conjugate with respect to matrix if Q is

$$\left(d^{(i)}\right)^T Q\left(d^{(j)}\right) = 0, \quad \forall i \neq j, \ i = 1,2,3...n, j = 1,2,3....n$$

Parallel Subspace Property

Consider a quadratic function with two variables:

$$q(x) = \left\{ a + b^T x + \frac{1}{2} x^T Q x, a \text{ is scalar, } b \text{ is vector, Q is } 2 \times 2 \text{ matrix} \right.$$

Let $y^{(1)}$ is solution to the problem: minimize $q(x^{(1)} + \lambda d)$

And $y^{(2)}$ is the solution to the problem: minimize $q(x^{(2)} + \lambda d)$

Then the direction $(y^{(2)} - y^{(1)})$ is conjugate to d. It equivalently means that $(y^{(2)} - y^{(1)})Q d = 0$

Algorithm:

Step 0: Set $x^{(0)}$– initial approximation; set $d^{(i)}$ as a set of n linearly independent directions.

Step 1: Perform unidirectional search along $d^{(1)}$ and take it to $d^{(n)}$.

Step 2: Use extended parallel subspace property to form new conjugate direction, d.

Step 3: If $\|d\|$ is smaller or search directions are linearly dependent, then terminate the process.

Else replace $d^{(j)} = d^{(j-1)} \ \forall j = n, \ n-1, \ n-2....$

$d^{(1)} = \dfrac{d}{\|d\|}$ and then go to step 1.

Example:

Find the minimum of $f(x_1, x_2) = 2x_1^3 + 4x_1 x_2^3 - 10x_1 x_2 + x_2^2$, $\quad x^{(0)} = [5, \ 2]^T$

Solution

Step 0: Set $d^{(1)} = \begin{bmatrix} 1 \\ 0 \end{bmatrix}, d^{(2)} = \begin{bmatrix} 0 \\ 1 \end{bmatrix}, x^{(0)} = \begin{bmatrix} 5 \\ 2 \end{bmatrix}$

Step 1: $f\left(x^{(0)}\right) = 314$, we need to find the minimum of $f\left(x^{(0)} + \lambda d^{(2)}\right)$

$$f\left(x^{(0)} + \lambda d^{(2)}\right) = f\left(\begin{bmatrix} 5 \\ 2 \end{bmatrix} + \lambda \begin{bmatrix} 0 \\ 1 \end{bmatrix}\right) = f\left(\begin{bmatrix} 5 \\ 2+\lambda \end{bmatrix}\right) = 20\lambda^3 + 121\lambda^2 + 229\lambda + 384$$

$$\frac{df}{d\lambda} = 60\lambda^2 + 242\lambda + 229 = 0, \quad \lambda = -1.5 \quad \left.\frac{d^2 f}{d\lambda^2}\right|_{d=-1.5} = 120\lambda + 242 = 62 > 0$$

$$x^{(1)} = x^{(0)} + (\lambda^*)d^{(2)} = \begin{bmatrix} 5 \\ 2 \end{bmatrix} - 1.5\begin{bmatrix} 0 \\ 1 \end{bmatrix} = \begin{bmatrix} 5 \\ 1.5 \end{bmatrix}$$

Step 2: $f\left(x^{(1)}\right) = 244.75$, we need to find the minimum of $f\left(x^{(1)} + \lambda d^{(1)}\right)$

$$f\left(x^{(1)} + \lambda d^{(1)}\right) = f\left(\begin{bmatrix} 5 \\ 1.5 \end{bmatrix} + \lambda \begin{bmatrix} 1 \\ 0 \end{bmatrix}\right) = f\left(\begin{bmatrix} 5+\lambda \\ 2 \end{bmatrix}\right) = 2\lambda^3 + 30\lambda^2 + 148.5\lambda - 5.25$$

$$\frac{df}{d\lambda} = 6\lambda^2 + 60\lambda + 148.5 = 0, \quad \lambda = -4.5, \quad \left.\frac{d^2 f}{d\lambda^2}\right|_{d=-4.5} = 12\lambda + 60 = 6 > 0$$

$$x^{(2)} = x^{(1)} + (\lambda^*)d^{(2)} = \begin{bmatrix} 5 \\ 1.5 \end{bmatrix} - 4.5\begin{bmatrix} 1 \\ 0 \end{bmatrix} = \begin{bmatrix} 0.5 \\ 1.5 \end{bmatrix}, \quad f\left(x^{(2)}\right) = 1.75.$$

2.2.5 SIMPLEX ALGORITHM

This is a popular search technique that uses $(n + 1)$ points for any problem with n decision variables. On the basis of functional value of these $(n + 1)$ points, a worst point is identified. Obviously, worst point is omitted and using rest of the points centroid is calculated. Further, new points are calculated on the basis of reflection, contraction, and expansion. Precise algorithm is mentioned below:

Algorithm:

Step 0: Set $(n + 1)$ points to define initial Simplex. Set $\alpha, \gamma > 1$, $\beta \in (0,1)$ also $\varepsilon > 0$ (for termination).

Step 1: Get the worst and best points from these $(n + 1)$ points. Let x_w and x_b be the worst and best points.

Step 2: Calculate the centroid x_c from the remaining (n) points using

$$x_c = \frac{1}{n}\sum_{\substack{i=1 \\ i \neq w}}^{n+1} x_k.$$

Step 3: Replace the worst point with new point as x_{new}, using x_c through the process of reflection by $x_r = (1+\alpha)x_w - \alpha x_b$.

(i) If $f(x_r) < f(x_b)$, then $x_{new} = (1+\gamma)x_c - \gamma x_w$
(ii) If $f(x_r) \geq f(x_w)$, then $x_{new} = (1-\beta)x_c - \beta x_w$
(iii) If $f(x_g) < f(x_r) < f(x_n)$, then $x_{new} = (1+\beta)x_c - \beta x_n$

Declare x_{new} as improved point

Step 4: If $\left[\sum\limits_{i=1}^{N+1} \dfrac{\left[f(x_i) - f(x_b) \right]^2}{N+1} \right]^{1/2} \leq \varepsilon$, then stop or else go to step 2

Example:

Minimize $f(x_1, x_2) = x_1 - x_2 + 2x_1^2 + 2x_1 x_2 + x_2^2$.

For initial simplex consider $X_1 = \begin{bmatrix} 4 \\ 4 \end{bmatrix}$, $X_2 = \begin{bmatrix} 5 \\ 4 \end{bmatrix}$, $X_3 = \begin{bmatrix} 4 \\ 5 \end{bmatrix}$ and $\alpha = 1.0$, $\beta = 0.5$

$\gamma = 2.0$, set $\varepsilon = 0.2$

Solution:

Iteration 1:

Step 1: Since there are two unknowns, initially we need three simplex as X_1, X_2 and X_3 $f(X_1) = 80, f(X_2) = 107, f(X_3) = 96$.

Then, $X_2 = \begin{bmatrix} 5 \\ 4 \end{bmatrix}$ is best whereas $X_1 = \begin{bmatrix} 4 \\ 4 \end{bmatrix}$ is worst point

Therefore, $X_b = \begin{bmatrix} 5 \\ 4 \end{bmatrix}$, $X_w = \begin{bmatrix} 4 \\ 4 \end{bmatrix}$

Step 2: The centroid X_c is $X_c = \dfrac{1}{2}(X_1 + X_3) = \begin{bmatrix} 4 \\ 5 \end{bmatrix}$, where $f(X_c) = 87.25$

Step 3: The reflection point is

$X_r = 2X_c - X_w = 2\begin{bmatrix} 4 \\ 4.5 \end{bmatrix} - \begin{bmatrix} 5 \\ 4 \end{bmatrix} = \begin{bmatrix} 3 \\ 5 \end{bmatrix}$, where $f(X_r) = 71$

Step 4: As $f(X_r) < f(X_b)$, we will go with expansion as

$X_e = 2X_r - X_c = 2\begin{bmatrix} 3 \\ 5 \end{bmatrix} - \begin{bmatrix} 4 \\ 4.5 \end{bmatrix} = \begin{bmatrix} 2 \\ 5.5 \end{bmatrix}$, where $f(X_e) = 56.75$

Step 5: $f(X_e) < f(X_b)$, we replace X_w by X_e

Step 6: Calculate Q for convergence which is 19.06, so we will continue with next iteration.

Iteration 2: We have $X_1 = \begin{bmatrix} 4 \\ 4 \end{bmatrix}$, $X_2 = \begin{bmatrix} 2 \\ 5.5 \end{bmatrix}$, $X_3 = \begin{bmatrix} 4 \\ 5 \end{bmatrix}$,

Step 1: $f(X_1) = 80$, $f(X_2) = 56.75$, $f(X_3) = 96$

$X_b = \begin{bmatrix} 2 \\ 5.5 \end{bmatrix}$, $X_w = \begin{bmatrix} 4 \\ 5 \end{bmatrix}$ are best and worst points respectively.

Step 2: The centroid X_c is $X_c = \dfrac{1}{2}(X_1 + X_2) = \begin{bmatrix} 3 \\ 4.75 \end{bmatrix}$, where $f(X_c) = 67.31$

Step 3: The reflection point is

$X_r = 2X_c - X_w = 2\begin{bmatrix} 3 \\ 4.75 \end{bmatrix} - \begin{bmatrix} 4 \\ 5 \end{bmatrix} = \begin{bmatrix} 2 \\ 4.5 \end{bmatrix}$, where $f(X_r) = 43.75$

Step 4: As $f(X_r) < f(X_b)$, we will go with expansion as

$X_e = 2X_r - X_c = 2\begin{bmatrix} 2 \\ 4.5 \end{bmatrix} - \begin{bmatrix} 3 \\ 4.75 \end{bmatrix} = \begin{bmatrix} 1 \\ 4.25 \end{bmatrix}$, where $f(X_e) = 23.3125$

Step 5: $f(X_e) < f(X_b)$, we replace X_w by X_e, we obtain the new vertices as

$X_1 = \begin{bmatrix} 4 \\ 4 \end{bmatrix}$, $X_2 = \begin{bmatrix} 2 \\ 5.5 \end{bmatrix}$, $X_3 = \begin{bmatrix} 1 \\ 4.25 \end{bmatrix}$

Step 6: For convergence, we compute $Q = 26.1 > \varepsilon$, we will go to next iteration.

2.3 GRADIENT-BASED METHODS

2.3.1 USING HESSIAN MATRIX

Let the objective function be *Min* $z = f(x) = f(x_1, x_2 .....x_n)$.
 Gradient vector of $f(x)$ is denoted by

$$\nabla f(x) = \left[\frac{\partial f(x)}{\partial x_1}, \frac{\partial f(x)}{\partial x_2} ....., \frac{\partial f(x)}{\partial x_n} \right]^T$$

$$H(x) = \begin{vmatrix} \dfrac{\partial^2 f(x)}{\partial x_1} & \dfrac{\partial^2 f(x)}{\partial x_1 \partial x_2} & \cdots & \dfrac{\partial^2 f(x)}{\partial x_1 \partial x_n} \\[2mm] \dfrac{\partial^2 f(x)}{\partial x_2 \partial x_1} & \dfrac{\partial^2 f(x)}{\partial x_2^2} & \cdots & \dfrac{\partial^2 f(x)}{\partial x_2 \partial x_n} \\[2mm] \vdots & \vdots & & \vdots \\[2mm] \dfrac{\partial^2 f(x)}{\partial x_n \partial x_1} & \dfrac{\partial^2 f(x)}{\partial x_n \partial x_2} & & \dfrac{\partial^2 f(x)}{\partial x_n^2} \end{vmatrix}^T$$

Necessary condition: For a continuous function $f(x)$ to have extreme point at $x = x_0$
is that the gradient $\nabla f(x_0) = 0$. That is $\dfrac{\partial f(x_0)}{\partial x_1} = \dfrac{\partial f(x_0)}{\partial x_2} = \dfrac{\partial f(x_0)}{\partial x_n} = 0$

Sufficient Condition: A stationary point $x = x_0$ is extreme point if Hessian matrix $H(x_0)$ is

- Positive definite when $x = x_0$ is a minimum point and
- Negative definite when $x = x_0$ is a maximum point.

Example:
Determine the maximum of the function $f(x_1, x_2) = x_1 + 2x_2 + x_1 x_2 - x_1^2 - x_2^2$

Solution
The necessary condition for local optimum value is that gradient

$$\nabla f(x) = \left[\frac{\partial f}{\partial x_1}, \frac{\partial f}{\partial x_2} \right] = \begin{bmatrix} 2x_1 x_2 + 5e^{x_2} \\ x_1^2 + 5x_1 e^{x_2} \end{bmatrix} = \begin{bmatrix} 0 \\ 0 \end{bmatrix}$$

Stationary point is $x_0 = \left[\dfrac{4}{3}, \dfrac{5}{3} \right]$

The sufficient condition using Hessian matrix is

$$H(x) = \begin{bmatrix} \dfrac{\partial^2 f}{\partial x_1^2} & \dfrac{\partial^2 f}{\partial x_1 \partial x_2} \\ \dfrac{\partial^2 f}{\partial x_2 \partial x_1} & \dfrac{\partial^2 f}{\partial x_2^2} \end{bmatrix} = \begin{bmatrix} -2 & 1 \\ 1 & -2 \end{bmatrix}$$

Since, $H(x)$ is negative definite the stationary point $x_0 = \left[\dfrac{4}{3}, \dfrac{5}{3} \right]$ is local maximum of the function $f(x)$.

2.3.2 STEEPEST DESCENT METHOD

This is a gradient-based method that uses gradient information to locate local minima of the function. In principle, functional value of the function always increases faster along the gradient direction. Hence, the functional value decreases faster in the negative direction of gradient. On the basis of this we can move to steepest ascent in the direction along the gradient if we need to achieve maxima and similarly to steepest descent in the negative direction of the gradient in case we need to achieve minima. Here, we have demonstrated algorithm as well as example for a minima which is steepest descent method. In general, we start with an initial guess and with a step length of α that move in the direction of $-\nabla f$.

Algorithm:

Step 0: Set $x^{(0)}$, $k = 0$
Step 1: $d^{(K)} = -\nabla f(x^{(K)})$. If $d^{(K)} = 0$, then stop.

Step 2: Solve $\min_\alpha f(x^{(K)} + \alpha^{(K)} d^{(K)})$ for step size $\alpha^{(K)}$ chosen by line search method.

Step 3: Set $x^{(K+1)} = x^{(K)} + \alpha^{(K)} d^{(K)}$, $K = K + 1$. Go to step 1.

Note: $d^{(K)} = -\nabla f(x^{(K)})$ is descent direction, it follows $f(x^{(K+1)}) < f(x^{(K)})$.

Example:

Determine the minimum of the given function $f(x_1, x_2) = x_1 - x_2 + 2x_1^2 + 2x_1 x_2 + x_2^2$ using steepest descent method with initial guess $x_0 = \begin{bmatrix} 0 \\ 0 \end{bmatrix}$.

Solution:

For $f(x_1, x_2) = x_1 - x_2 + 2x_1^2 + 2x_1 x_2 + x_2^2$, $x_0 = \begin{bmatrix} 0 \\ 0 \end{bmatrix}$ and $\nabla f = \begin{bmatrix} 1 + 4x_1 + 2x_2 \\ -1 + 2x_1 + 2x_2 \end{bmatrix}$

First iteration:

We have, $x_0 = \begin{bmatrix} 0 \\ 0 \end{bmatrix}$, $\nabla f|_{x_0} = \begin{bmatrix} 1 \\ -1 \end{bmatrix} \Rightarrow d^{(0)} = \begin{bmatrix} -1 \\ 1 \end{bmatrix}$,

$\min_\alpha f(x^{(0)} + \alpha d^{(0)}) = \min_\alpha f\left(\begin{bmatrix} 0 \\ 0 \end{bmatrix} + \alpha^{(0)} \begin{bmatrix} -1 \\ 1 \end{bmatrix} \right) = \min_\alpha f\left(\begin{bmatrix} -\alpha \\ \alpha \end{bmatrix} \right) = \alpha^2 - 2\alpha$,

$\dfrac{df}{d\alpha} = 0 \Rightarrow \alpha = 1$. Now generate $x^{(1)}$ by $x^{(K+1)} = x^{(K)} + \alpha \, d^{(K)}$ taking k = 0

$x^{(1)} = x^{(0)} + \alpha \, d^{(0)} = \begin{bmatrix} 0 \\ 0 \end{bmatrix} + 1 \begin{bmatrix} -1 \\ 1 \end{bmatrix} = \begin{bmatrix} -1 \\ 1 \end{bmatrix}$. Since, $\nabla f|_{X^{(1)}} = \begin{bmatrix} -1 \\ -1 \end{bmatrix} \neq 0$. We will go for next iteration.

Second Iteration: $x_1 = \begin{bmatrix} -1 \\ -1 \end{bmatrix}$ and $\nabla f|_{X^{(1)}} = \begin{bmatrix} -1 \\ -1 \end{bmatrix} \Rightarrow d^{(1)} = \begin{bmatrix} 1 \\ 1 \end{bmatrix}$

Optimal direction is $\alpha = 0.2$, $x^{(2)} = x^{(1)} + \alpha \, d^{(1)} = \begin{bmatrix} -1 \\ 1 \end{bmatrix} + 0.2 \begin{bmatrix} 1 \\ 1 \end{bmatrix} = \begin{bmatrix} -0.8 \\ 1.2 \end{bmatrix}$

Since $\nabla f|_{X^{(2)}} = \begin{bmatrix} 0.2 \\ -0.2 \end{bmatrix} \neq 0$. We will continue the process till desired accuracy is achieved.

2.3.3 NEWTON'S METHOD

This method approximates the given function by second-order Taylor's approximation. Further, that approximation is optimized using necessary and sufficient condition of calculus for optimal value.

$$f(x) = f(x_0) + \nabla f(x_0)h + \frac{1}{2!} h^T H(x)h, \quad x = x_0 + h$$

It can be represented in standard form as $q(x) = \dfrac{1}{2} x^T Hx + b^T x + c$

Necessary condition: $\nabla q(x) = 0 \Rightarrow Hx + b = 0 \Rightarrow x = -H^{-1}b$, where $b = \nabla f(x)$
Sufficient condition: $\nabla^2 q(x) = 0 \Rightarrow H = 0 \Rightarrow f(x)$ is Minimum at $x = x_0$ if H is positive definite.

Algorithm:

Step 0: Set $x_0 \in R^n, k = 0$, ε(very small quantity)
Step 1: Find $x_{K+1} = x_K + H_K^{-1}\nabla f(x_k)$
Step 2: If $\nabla f(x_k) < \varepsilon$, terminate the process otherwise go to step 1

Example:
Determine the minimum of the given function $f(x_1, x_2) = x_1 - x_2 + 2x_1^2 + 2x_1x_2 + x_2^2$
using Newton's method with initial guess $x_0 = \begin{bmatrix} 0 \\ 0 \end{bmatrix}$

Solution:
For the given $f(x_1, x_2) = x_1 - x_2 + 2x_1^2 + 2x_1x_2 + x_2^2$

$\dfrac{\partial f}{\partial x_1} = 1 + 4x_1 + 2x_2, \quad \dfrac{\partial^2 f}{\partial x_1^2} = 4, \quad \dfrac{\partial f}{\partial x_2} = -1 + 2x_1 + 2x_2, \quad \dfrac{\partial^2 f}{\partial x_2^2} = 2$

First iteration: $x_0 = \begin{bmatrix} 0 \\ 0 \end{bmatrix}$

$$H_0 = \begin{bmatrix} 4 & 2 \\ 2 & 2 \end{bmatrix}, H_0^{-1} = \dfrac{1}{4}\begin{bmatrix} 2 & -2 \\ -2 & 4 \end{bmatrix} = \begin{bmatrix} \dfrac{1}{2} & -\dfrac{1}{2} \\ -\dfrac{1}{2} & 1 \end{bmatrix} \text{ and } \nabla f_0 = \begin{bmatrix} -1 \\ -1 \end{bmatrix}$$

$$x_1 = x_0 + H_0^{-1}\nabla f(x_0) = \begin{bmatrix} 0 \\ 0 \end{bmatrix} - \begin{bmatrix} \dfrac{1}{2} & -\dfrac{1}{2} \\ -\dfrac{1}{2} & 1 \end{bmatrix}\begin{bmatrix} 1 \\ -1 \end{bmatrix} = \begin{bmatrix} 0 \\ 0 \end{bmatrix} - \begin{bmatrix} 1 \\ -3/2 \end{bmatrix} = \begin{bmatrix} -1 \\ 3/2 \end{bmatrix}$$

Now, $\nabla f|_{x_1} = \begin{bmatrix} 0 \\ 0 \end{bmatrix}$.

We can terminate the process here since gradient is 0. Thus minima is $\begin{bmatrix} -1 \\ 3/2 \end{bmatrix}$

2.3.4 QUASI METHOD

This method is a good alternative of Newton's method. It computes search direction only by first derivatives whereas Newton's method uses Hessian matrix for the same purpose. This method basically uses approximation of Hessian matrix and thus possess less computation cost compared to Newton's method. This approximation is initialized as a positive definite matrix and is updated on the basis of previous points and gradients.

Algorithm:

Step 1: Compute Newton direction as $d^k = \tilde{H}_k g^k$, where $\tilde{H}_k$ and g^k are approximate inverse Hessian matrix and gradient, respectively.

Step 2: Compute new approximation as $x^{k+1} = x^k + \alpha_k d^k$

Step 3: Compute $g^{k+1} = \nabla f(x^{k+1})$

Step 4: Update approximate inverse Hessian matrix using
$$\tilde{H}_{k+1} = update(\tilde{H}_k, x^{k+1} \pm x^k, g^{k+1} - g^k)$$

Calculation of approximated Hessian matrix

Let the difference of two successive $x^{(s)}$ as $p^{(k)} = x^{(k+1)} - x^{(k)}$

Let difference of two successive gradients as $q^{(k)} = g^{(k+1)} - g^{(k)}$

As per the gradient and Hessian matrix relation $Hdx = dg \Rightarrow Hp^k = q^k$ is known as secant equation.

Then, rank one update is $H_{k+1} = H_k + uv^T$, where u and v are collinear and symmetric.

$$\Rightarrow u(v^T p^k) = q^k - H_k p^k \Rightarrow u = \frac{1}{v^T p^k}(q^k - H_k p^k) \Rightarrow \frac{1}{v^T p^k} v,$$

where $v = q^k - H_k p^k$, $H_{k+1} = H_k - \frac{1}{v^T p^k} vv^T$

Sherman–Morrison Formula: $\left(A + uv^T\right)^{-1} = A^{-1} + \frac{1}{1 - v^T A^{-1} u} A^{-1} uv^T A^{-1}$

To update the Hessian matrix $Min \left\| \tilde{H}^{k+1} - \tilde{H}^k \right\|_w$ and $\tilde{H}^{k+1} > 0$

$W \approx H \Leftarrow$ broyden-fletcher-goldfarb-shanno

$W \approx H^{-1} \Leftarrow$ davidon fletcher powell method

These two approaches are widely used to update the Hessian matrix approximation.

TRY YOURSELF

Q1. Find the minimum for $f(x_1, x_2) = 3x_1^2 + x_2^2 - 10$ using the following methods with [0, 0] initial approximation [0, 0] using Univariate Method and Hooke–Jeeves Method

Answer: [6, 2]

Q2. Minimize the objective function $f(x, y) = x - y - 2x^2 + 2xy + y^2$ using 5 iterations of (a) Newton's method (b) Steepest Descent with starting value $x_0 = [0,0]^T$. Plot the values of iterates for each method on the same graph. Which method seems to be more efficient.

Answer: [−1, 1.5]

Q3. Minimize $f(x,y) = x^4 - 2x^2y + x^2 + y^2 + 2x + 1$ by the Simplex method. Perform two steps of reflection, expansion, and/or contraction.
Answer: [1, 1]

Q4. Minimize $f(x,y) = 4x^2 - 3y^2 - 5xy - 8x$ starting from point [0, 0] using Powell's method. Perform four iterations.
Answer: [2.0869, 1.7390]

3 Constrained Multivariable Optimization

3.1 INTRODUCTION

We have already discussed methods for optimization of single variable and multivariable without constraint in Chapter 1 and Chapter 2, respectively. Actually, most of the real world problems that are required to be optimized have constraints. Now, it is high time to discuss optimization of multivariable functions with constraints. General structure of multivariable functions with constraints is given below:

$$\text{Optimize (max or min) } Z = f(x_1, x_2, ..., x_n)$$

subject to the constraints

$$h_i(x_1, x_2, ..., x_n) \leq = \geq b_i; \quad i = 1, 2, ..., m$$

This set of problems can be further divided into problem with equality constraints and unequality constraints. There are some conventional methods available for both set of problems with and without constraints but still all the problems cannot be solved using these methods due to complexity of the problem. Some of the time, conventional methods stuck to the local optimum instead of global optimum. So, we have another set of methods known as stochastic search techniques. These methods are search algorithms inspired by natural phenomenon like evolution, natural selection, animal behaviour, or natural laws. Under this section, we will discuss methods like genetic algorithm, particle swarm optimization, simulated annealing, and Tabu search. For the sake of convenience, first we will discuss the conventional methods then followed by stochastic search techniques.

3.2 CONVENTIONAL METHODS FOR CONSTRAINED MULTIVARIATE OPTIMIZATION

3.2.1 PROBLEMS WITH EQUALITY CONSTRAINTS

Optimize (max or min)$Z = f(x_1, x_2, ..., x_n)$ subject to the constraints

$$h_i(x_1, x_2, ..., x_n) = b_i; i = 1, 2, ..., m$$

In matrix notation the above problem can also be written as:

$$Z = f(x)$$

subject to the constraints

$$g_i(x) = 0, i = 1, 2, ..., m \quad \text{where } x = (x_1, x_2, ..., x_n),$$

and $g_i(x) = h_i(x) - b_i$; b_i is constant

Here it is assumed that $m < n$ to get the solution.

There are various methods for solving the above defined problem. But in this section, we shall discuss only two methods: (i) Direct Substitution Method and (ii) Lagrange Multiplier Method.

3.2.1.1   Direct Substitution Method

Example:

Find the optimum solution of the following constrained multivariable problem:

$$\text{Minimize } Z = x_1^2 + (x_2 + 1)^2 + (x_3 - 1)^2$$

subject to the constraint

$$x_1 + 5x_2 - 3x_3 = 6$$

and $x_1, x_2, x_3 \geq 0$

Solution

Since the given problem has three variables and one equality constraint, any one of the variables can be removed from Z with the help of the equality constraint. Let us choose variable x_3 to be eliminated from Z. Then, from the equality constraint, we have:

$$x_3 = \frac{(x_1 + 5x_2 - 6)}{3}$$

Substituting the value of x_3 in the objective function, we get:

$$Z \text{ or } f(x) = x_1^2 + (x_2 + 1)^2 + \frac{1}{9}(x_1 + 5x_2 - 9)^2$$

The necessary condition for minimum of Z is that the gradient

$$\nabla f(x) = \left[\frac{\partial f}{\partial x_1}, \frac{\partial f}{\partial x_2} \right] = 0$$

That is, $\dfrac{\partial Z}{\partial x_1} = 2x_1 + \dfrac{2}{9}(x_1 + 5x_2 - 9) = 0$

$\dfrac{\partial Z}{\partial x_2} = 2(x_2 + 1) + \dfrac{10}{9}(x_1 + 5x_2 - 9) = 0$

On solving these equations, we get $x_1 = 2/5$ and $x_2 = 1$.

To find whether the solution, so obtained, is minimum or not, we must apply the sufficiency condition forming a Hessian matrix. The Hessian matrix for the given objective function is $H(x_1, x_2) = \begin{pmatrix} \dfrac{\partial^2 Z}{\partial x_1{}^2} & \dfrac{\partial^2 Z}{\partial x_1 \partial x_2} \\ \dfrac{\partial^2 Z}{\partial x_2 \partial x_1} & \dfrac{\partial^2 Z}{\partial x_2{}^2} \end{pmatrix} = \begin{pmatrix} 20/9 & 10/9 \\ 10/9 & 20/9 \end{pmatrix}$

Since the matrix is symmetric and principal diagonal elements are positive, $H(x_1, x_2)$ is positive definite and the objective function is convex. Hence, the optimum solution of the given problem is $x_1 = 2/5$, $x_2 = 1$, $x_3 = -1/5$, and Min $Z = 28/5$.

3.2.1.2 Lagrange Multipliers Method

Optimize $Z = f(x)$

subject to the constraint

$$h_i(x) = b_i$$

or

$$g_i(x) = h_i(x) - b_i = 0, \ i = 1, 2, ..., m \text{ and } m \le n \ ; x \in E^n$$

The necessary conditions for a function to have a local optimum at the given points can be extended to the case of a general problem with n variables and m equality constraints.

Multiply each constraint with an unknown variable $\lambda_i (i = 1, 2, ..., m)$ and subtract each from the objective function $f(x)$ to be optimized. The new objective function now becomes:

$$L(x, \lambda) = f(x) - \sum_{i=1}^{m} \lambda_i g_i(x) \ ; \ x = (x_1, x_2, ..., x_n)^T$$

where $m < n$. The function $L(x, \lambda)$ is called the *Lagrange function*.

The necessary conditions for an unconstrained optimum of $L(x, \lambda)$, i.e. the first derivatives, with respect to x and λ of $L(x, \lambda)$ must be zero, are also necessary conditions for the given constrained optimum of $f(x)$, provided the matrix of partial derivatives $\partial g_i / \partial x_j$ has rank m at the point of optimum.

The necessary conditions for an optimum (max or min) of $L(x, \lambda)$ or $f(x)$ are the $m + n$ equations to be solved for $m + n$ unknown $(x_1, x_2, ..., x_n; \lambda_1, \lambda_2, ..., \lambda_m)$.

$$\frac{\partial L}{\partial x_j} = \frac{\partial f}{\partial x_j} - \sum_{i=1}^{m} \lambda_i \frac{\partial g_i}{\partial x_j} = 0; \quad j = 1, 2, ..., n$$

$$\frac{\partial L}{\partial \lambda_i} = -g_i; \quad i = 1, 2, ..., m$$

These $m + n$ necessary conditions also become sufficient conditions for a maximum (or minimum) of the objective function $f(x)$, in case it is concave (or convex) and the constraints are equalities, respectively.

Sufficient conditions for a general problem

Let the Lagrangian for a general non-linear programming (NLP) problem, involving n variables and m $(< n)$ constraints, be

$$L(x, \lambda) = f(x) - \sum_{i=1}^{m} \lambda_i g_i(x)$$

Further, the necessary conditions

$$\frac{\partial L}{\partial x_j} = 0 \text{ and } \frac{\partial L}{\partial \lambda_i} = 0; \text{ for all } i \text{ and } j$$

For an extreme point to be local optimum of $f(x)$ is also true for optimum of $L(x, \lambda)$.

Let there exists points x and λ that satisfy the equation

$$\nabla L(x, \lambda) = \nabla f(x) - \sum_{i=1}^{m} \lambda_i g_i(x) = 0$$

and $g_i(x) = 0$, $i = 1, 2, ..., m$

Then the sufficient condition for an extreme point x to be a local minimum (or local maximum) of $f(x)$ subject to the constraints $g_i(x) = 0$, $(i = 1, 2, ..., m)$ is that the determinant of the matrix (*also called Bordered Hessian matrix*)

$$D = \begin{pmatrix} \mathbf{Q} & \mathbf{H} \\ \mathbf{H}^T & \mathbf{0} \end{pmatrix}_{(m+n) \times (m+n)}$$

is positive (or negative), where

$$Q = \left[\frac{\partial^2 L(x,\lambda)}{\partial x_i \partial x_j}\right]_{n \times n} \quad ; \; H = \left[\frac{\partial g_i(x)}{\partial x_j}\right]_{m \times n}$$

Conditions for maxima and minima:
The sufficient condition for the maxima and minima is determined by the signs of the last $(n - m)$ principal minors of matrix D. That is,

1. If starting with principal minor of order $(m+1)$, the extreme point gives the maximum value of the objective function when signs of last $(n - m)$ principal minors alternate, starting with the $(-1)^{m+n}$ sign.
2. If starting with principal minor of order $(2m+1)$ the extreme point gives the minimum value of the objective function when all signs of last $(n - m)$ principal minors are the same and are of $(-1)^m$ type.

Example:
Solve the following problem by using the method of Lagrangian multipliers.

$$\text{Minimize } Z = x_1^2 + x_2^2 + x_3^2$$

subject to the constraints

$$(i) \; x_1 + x_2 + 3x_3 = 2, \; (ii) \; 5x_1 + 2x_2 + x_3 = 5$$

and $x_1, x_2 \geq 0$

Solution:
The Lagrangian function is

$$L(x,\lambda) = x_1^2 + x_2^2 + x_3^2 - \lambda_1(x_1 + x_2 + 3x_3 - 2) - \lambda_2(5x_1 + 2x_2 + x_3 - 5)$$

The necessary conditions for the minimum of Z give us the following:
    The solution of these simultaneous equations gives:

$x = (x_1, x_2, x_3) = (37/46, 16/46, 13/46); \; \lambda = (\lambda_1, \lambda_2) = (2/23, 7/23)$ and $Z = 193/250$

To see that this solution corresponds to the minimum of Z, apply the sufficient condition with the help of a matrix:

$$D = \begin{bmatrix} 2 & 0 & 0 & 1 & 5 \\ 0 & 2 & 0 & 1 & 2 \\ 0 & 0 & 2 & 3 & 1 \\ 1 & 1 & 3 & 0 & 0 \\ 5 & 2 & 1 & 0 & 0 \end{bmatrix}$$

Since $m = 2, n = 3$, so $n - m = 1$ and $2m + 1 = 5$, only one of minor of D of order 5 needs to be evaluated and it must have a positive sign; $(-1)^m = (-1)^2 = 1$. Since $|D| = 460 > 0$, the extreme point, (x_1, x_2, x_3) = corresponds to the minimum of Z.

Necessary and sufficient conditions when concavity (convexity) of objective function is not known, with single equality constraint:

Let us consider the non-linear programming problem that involves n decision variables and a single constraint.

Optimize $Z = g(x)$

subject to the constraint

$$g(x) = h(x) - b = 0; \ x = (x_1, x_2, ..., x_n)^T \geq 0$$

Multiply each constraint by Lagrange multiplier λ and subtract it from the objective function. The new unconstrained objective function (Lagrange function) becomes:

$$L(x, \lambda) = f(x) - \lambda g(x)$$

The necessary conditions for an extreme point to be optimum (max or min) point are:

$$\frac{\partial L}{\partial x_j} = \frac{\partial f}{\partial x_j} - \lambda \frac{\partial g}{\partial x_j} = 0; \ j = 1, 2, ..., n$$

$$\frac{\partial L}{\partial \lambda} = -g(x) = 0$$

From the first condition we obtain the value of λ as:

$$\lambda = \frac{(\partial f / \partial x_j)}{(\partial g / \partial x_j)}; \ j = 1, 2, ..., n$$

The sufficient conditions for determining whether the optimal solution, so obtained, is either maximum or minimum, need computation of the value of $(n-1)$ principal minors, of the determinant, for each extreme point as follows:

$$\Delta_{n+1} = \begin{vmatrix} 0 & \dfrac{\partial g}{\partial x_1} & \dfrac{\partial g}{\partial x_2} & \cdots & \dfrac{\partial g}{\partial x_n} \\[2mm] \dfrac{\partial g}{\partial x_1} & \dfrac{\partial^2 f}{\partial x_1^2} - \lambda \dfrac{\partial^2 g}{\partial x_1^2} & \dfrac{\partial^2 f}{\partial x_1 \partial x_2} - \lambda \dfrac{\partial^2 g}{\partial x_1 \partial x_2} & \cdots & \dfrac{\partial^2 f}{\partial x_1 \partial x_n} - \lambda \dfrac{\partial^2 g}{\partial x_1 \partial x_n} \\[2mm] \dfrac{\partial g}{\partial x_2} & \dfrac{\partial^2 f}{\partial x_2 \partial x_1} - \lambda \dfrac{\partial^2 g}{\partial x_2 \partial x_1} & \dfrac{\partial^2 f}{\partial x_2^2} - \lambda \dfrac{\partial^2 g}{\partial x_2^2} & \cdots & \dfrac{\partial^2 f}{\partial x_2 \partial x_n} - \lambda \dfrac{\partial^2 g}{\partial x_2 \partial x_n} \\[2mm] \vdots & \vdots & \vdots & & \vdots \\[2mm] \dfrac{\partial g}{\partial x_n} & \dfrac{\partial^2 f}{\partial x_n \partial x_1} - \lambda \dfrac{\partial^2 g}{\partial x_n \partial x_1} & \dfrac{\partial^2 f}{\partial x_n \partial x_2} - \lambda \dfrac{\partial^2 g}{\partial x_n \partial x_2} & \cdots & \dfrac{\partial^2 f}{\partial x_n^2} - \lambda \dfrac{\partial^2 g}{\partial x_n^2} \end{vmatrix}$$

If the sign of minors Δ_3, Δ_4, Δ_5 are alternately positive and negative, then the extreme point is a local maximum. But if sign of all minors Δ_3, Δ_4, Δ_5 are negative, then the extreme point is local minimum.

Example:
Use the method of Lagrangian multipliers to solve the following NLP problem. Does the solution maximize or minimize the objective function?

Optimize $Z = 2x_1^2 + x_2^2 + 3x_3^2 + 10x_1 + 8x_2 + 6x_3 - 100$

subject to the constraint

$$g(x) = x_1 + x_2 + x_3 = 20$$

and

$$x_1, x_2, x_3 \geq 0$$

Solution
Lagrangian function can be formulated as:

$$L(x, \lambda) = 2x_1^2 + x_2^2 + 3x_3^2 + 10x_1 + 8x_2 + 6x_3 - 100 - \lambda(x_1 + x_2 + x_3 - 20)$$

The necessary conditions for maximum or minimum are:

$$\frac{\partial L}{\partial x_1} = 4x_1 + 10 - \lambda = 0; \quad \frac{\partial L}{\partial x_2} = 2x_2 + 8 - \lambda = 0$$

$$\frac{\partial L}{\partial x_3} = 6x_3 + 6 - \lambda = 0; \quad \frac{\partial L}{\partial \lambda} = -(x_1 + x_2 + x_3 - 20) = 0$$

Putting the values of x_1, x_2 and x_3 in the last equation $\partial L / \partial \lambda = 0$ and solving for λ, we get $\lambda = 30$. Substituting the value of λ in the other three equations, we get an extreme point: $(x_1, x_2, x_3) = (5, 11, 4)$.

To prove the sufficient condition of whether the extreme point solution gives maximum or minimum value of the objective function we evaluate $(n-1)$ principal minors as follows:

$$\Delta_3 = \begin{vmatrix} 0 & \dfrac{\partial g}{\partial x_1} & \dfrac{\partial g}{\partial x_2} \\[2mm] \dfrac{\partial g}{\partial x_1} & \dfrac{\partial^2 f}{\partial x_1^2} - \lambda \dfrac{\partial^2 g}{\partial x_1^2} & \dfrac{\partial^2 f}{\partial x_1 \partial x_2} - \lambda \dfrac{\partial^2 g}{\partial x_1 \partial x_2} \\[2mm] \dfrac{\partial g}{\partial x_2} & \dfrac{\partial^2 f}{\partial x_2 \partial x_1} - \lambda \dfrac{\partial^2 g}{\partial x_2 \partial x_1} & \dfrac{\partial^2 f}{\partial x_2^2} - \lambda \dfrac{\partial^2 g}{\partial x_2^2} \end{vmatrix} = \begin{vmatrix} 0 & 1 & 1 \\ 1 & 4 & 0 \\ 1 & 0 & 2 \end{vmatrix} = -6$$

$$\Delta_4 = \begin{vmatrix} 0 & 1 & 1 & 1 \\ 1 & 4 & 0 & 0 \\ 1 & 0 & 2 & 0 \\ 1 & 0 & 0 & 6 \end{vmatrix} = 48$$

Since, the sign of Δ_3 and Δ_4 are alternative, therefore extreme points: $(x_1, x_2, x_3) = (5, 11, 4)$ is local maximum. At this point the value of the objective function is $Z = 281$.

3.2.2   PROBLEMS WITH INEQUALITY CONSTRAINTS

3.2.2.1   Kuhn–Tucker Necessary Conditions

$$\text{Optimize } Z = f(x)$$

subject to the constraints

$$g_i(x) = h_i(x) - b_i \leq 0 \; i = 1, 2, ..., m \quad where \; x = (x_1, x_2, ..., x_n)^T$$

Add non-negative slack variables s_i $(i = 1, 2, ..., m)$ in each of the constraints to convert them to equality constraints. The problem can then be restated as:

$$\text{Optimize } Z = g(x)$$

subject to the constraints

$$g_i(x) + s_i^2 = 0, \; i = 1, 2, ..., m$$

The s_i^2 has only been added to ensure non-negative (feasibility requirement) of s_i and to avoid adding $s_i \geq 0$ as an additional side constraint. The new problem is the constrained multivariable optimization problem with equality constraints with $n + m$ variables. Thus, it can be solved using Lagrangian multiplier method. For this, let us form the Lagrangian function as:

$$L(x, s, \lambda) = f(x) - \sum_{i=1}^{m} \lambda_i [g_i(x) + s_i^2]$$

where $\lambda = (\lambda_1, \lambda_2, ..., \lambda_n)^T$ is the vector of Lagrange multiplier.

The Kuhn–Tucker necessary conditions (when active constraints are known) to be satisfied at local optimum (max or min) point can be stated as follows:

$$\frac{\partial f}{\partial x_j} - \sum_{i=1}^{m} \lambda_i \frac{\partial g_i}{\partial x_j} = 0, \; j = 1, 2, ..., n$$

$$\lambda_i g_i(x) = 0,$$

$$g_i(x) \le 0,$$

$$\lambda_i \ge 0, \ i = 1,2,...,m$$

3.2.2.2   Kuhn–Tucker Sufficient Conditions
The Kuhn–Tucker necessary conditions for the problem

$$\text{Maximize } Z = f(x)$$

subject to the constraints

$$g_i(x) \le 0, \ i = 1,2,...,m$$

are also sufficient conditions if $f(x)$ is concave and $g_i(x)$ are convex functions of x.

Example:
Maximize $Z = 12x_1 + 21x_2 + 21x_1x_2 - 2x_1^2 - 2x_2^2$ subject to the constraints

$$(i) \ x_2 \le 8, \ (ii) \ x_1 + x_2 \le 10,$$

and

$$x_1, x_2 \ge 0$$

Solution
Here $f(x_1, x_2) = 12x_1 + 21x_2 + 21x_1x_2 - 2x_1^2 - 2x_2^2$

$$g_1(x_1, x_2) = x_2 - 8 \le 0$$
$$g_2(x_1, x_2) = x_1 + x_2 - 10 \le 0$$

The Lagrangian function can be formulated as:

$$L(x,s,\lambda) = f(x) - \lambda_1[g_1(x) + s_1^2] - \lambda_2[g_2(x) + s_2^2]$$

The Kuhn–Tucker necessary condition can be stated as:

$(i) \ \dfrac{\partial f}{\partial x_j} - \displaystyle\sum_{i=1}^{2} \lambda_i \dfrac{\partial g_i}{\partial x_j} = 0, \ j = 1,2 \qquad (ii) \ \lambda_i g_i(x) = 0, \ i = 1.2$

$12 + 2x_2 - 4x_1 - \lambda_2 = 0 \qquad\qquad \lambda_1(x_2 - 8) = 0$

$21 + 2x_1 - 4x_2 - \lambda_1 - \lambda_2 = 0 \qquad\quad \lambda_2(x_1 + x_2 - 10) = 0$

$$(iii)\ g_i(x) \leq 0 \qquad\qquad\qquad (iv)\ \lambda_i \geq 0.\ i = 1,2$$

$$x_2 - 8 \leq 0$$

$$x_1 + x_2 - 10 \leq 0$$

There may arise four cases:

Case 1: If $\lambda_1 = 0$, $\lambda_2 = 0$, then from condition (i) we have:

$$12 + 2x_2 - 4x_1 = 0 \text{ and } 21 + 2x_1 - 4x_2 = 0$$

Solving these equations, we get $x_1 = 15/2$, $x_2 = 9$. However, this solution violates condition (iii) and so it should be discarded.

Case 2: $\lambda_1 \neq 0$, $\lambda_2 \neq 0$, then from condition (ii) we have:

$$x_2 - 8 = 0 \text{ or } x_2 = 8$$

$$x_1 + x_2 - 10 = 0 \text{ or } x_1 = 2$$

Substituting these values in condition (i), we get $\lambda_1 = -27$ and $\lambda_2 = 20$. However, this solution violates condition (iv) and therefore may be discarded.

Case 3: $\lambda_1 \neq 0$, $\lambda_2 = 0$, then from conditions (i) and (ii) we have:

$$x_1 + x_2 = 10$$

$$2x_2 - 4x_1 = -12$$

$$2x_1 - 4x_2 = -12 + \lambda_1$$

Solving these equations, we get $x_1 = 17/4$, $x_2 = 23/4$, and $\lambda_1 = -16$. However, this solution violates condition (iv) and therefore may be discarded.

Case 4: $\lambda_1 = 0$, $\lambda_2 \neq 0$, then from conditions (i) and (ii) we have:

$$2x_2 - 4x_1 = -12 + \lambda_2$$

$$2x_1 - 4x_2 = -21 + \lambda_2$$

$$x_1 + x_2 = 10$$

Solving these equations, we get

$x_1 = 17/4$, $x_2 = 23/4$, and $\lambda_2 = 13/4$. This solution does not violate any of the Kuhn–Tucker conditions and therefore must be accepted. Hence, the optimum solution of the given problem is $x_1 = 17/4$, $x_2 = 23/4$, $\lambda_1 = 0$, and $\lambda_2 = 13/4$. and Max $Z = 1734/16$.

3.3 STOCHASTIC SEARCH TECHNIQUES

Many search algorithms are available and are continuously being developed on the basis of different natural phenomenon. We shall discuss some of them to make our readers understand the basic idea or inspiration behind them. These methods can be used for problems with both equal and unequal constraints.

3.3.1 GENETIC ALGORITHM

Genetic algorithm (GA) is a search technique inspired by Charles Darwin's theory of "Survival of fittest". It is based on the natural selection process that allows only good chromosomes to go in the next generation to improve the chance of survival. John Holland introduced genetic algorithms in 1960 based on the concept of Darwin's theory of evolution, and his student David E. Goldberg further extended GA in 1989.

GA is a search technique used in computing to find true or approximate solutions for optimization of any non-linear problem. GA may explore the solution space in many directions and from many points using parallel computation process. Complex environments with non-lineal behaviour are good candidates to be worked with GA's. The fitness function may be discontinuous and even changing over time. Genetic algorithms are a particular class of evolutionary algorithms that use techniques inspired by evolutionary biology such as inheritance, mutation, selection, and crossover (also called recombination).

Fundamentals of Genetic algorithm
Initially, a population is created randomly with a group of individuals. These individuals are being evaluated on the basis of fitness function (objective function). A fitness function is defined by programmer over the genetic representation and measures the *quality* of the represented solution. The programmer provides score to individuals on the basis of their performance. Two individuals are selected on the basis of their fitness. Higher fitness score increases the chances of selection. Further, these best individuals reproduce offspring that are further muted on a random basis. This process continues until a feasible solution is attained. Let us discuss all the stages of GA step-by-step.

Initialization

- Initially many solutions are randomly generated to form an initial population. The population size depends on the nature of the problem, but typically contains several hundreds or thousands of possible solutions.
- Traditionally, the population is generated randomly, covering the entire range of possible solutions called *search space*.
- Sometimes, the solutions may be "seeded" i.e., selected from areas where there is possibility of getting optimal solutions.

Selection

- During each successive generation, a proportion of the present population is selected on the basis of fitness function to produce a new generation.
- Individual solutions are selected through a *fitness-based* process, where best fit solutions (as measured by a fitness function) are typically more likely to be selected. Some selection methods rate the fitness of each solution and prefer to select the best possible solutions. Other methods choose a random sample of the population, which can be more tedious.
- Due to stochastic approach of GA, even the less fit solutions are included in very small amount to diversify the population and preventing premature convergence to poor solutions. Roulette wheel selection and tournament selection are the widely used selection methods.

Reproduction

- A mating pool is created from the appropriate individuals for reproduction process. Members of mating pool crossbreed to generate new population. This approach is used to generate a next generation population of solutions from those selected through genetic operators: Crossover (also called recombination), and/or mutation.
- For each new solution to be generated, a pair of "parent" solutions is selected for breeding to get "child" solutions.
- By generating a "child" solution from either crossover or mutation, a new solution is created which typically shares many of the characteristics of its "parents". Now, new parents are selected for each child, and this process continues till a feasible solution set of appropriate size is generated.

3.3.1.1 Crossover

Parent

Parent 1	1	1	0	0	1	0	1	0	0	1
Parent 2	0	1	0	1	0	1	0	1	0	1

Two possible offsprings

Child 1	1	1	0	0	1	1	0	1	0	1
Child 2	0	1	0	1	0	0	1	0	0	1

Mutation

- After selection and crossover, we have a new population full of individuals in which some are simply copied and others are being crossover from parent chromosomes.
- To ensure that the individuals are not all exactly the same, we perform mutation.
- For this we examine the alleles of all the individuals, and if that allele is selected for mutation it can be changed by a small proportion or replaced with new value. The probability of a mutation of a bit is $1/L$, where L is the length of the binary vector.
- Mutation is fairly simple. You just change the selected alleles based on what you feel is necessary and move on. Mutation ensures genetic diversity within the population.

Before mutation	1	1	0	0	1	1	0	1	0	1
After mutation	1	1	0	0	0	1	0	1	0	1

It can also be flipped as follows

Before mutation	1	1	0	0	1	1	0	1	0	1
After mutation	0	0	1	1	0	0	1	0	1	0

Termination

- Above-discussed process gets repeated until a termination condition has been attained.
- Common terminating conditions are as follows:
- Manual inspection
- A solution is found that satisfies minimum criteria
- Fixed number of generations reached
- Allocated budget (computation time/money) reached
- The highest ranking solution's fitness is reaching or has reached a plateau such that successive iterations no longer produce better results
- Any combinations of the above

Limitations

- The representation of the problem may be difficult. You need to identify which variables are suitable to be treated as genes and which variables must be let outside the GA process.
- The determination of the convenient parameters (population size, mutation rate) may be time consuming
- As in any optimization process, if you don't take enough precautions, the algorithm may converge in a local minimum (or maximum).

3.3.2 PARTICLE SWARM OPTIMIZATION

Particle swarm optimization (PSO) is a population-based heuristic global search algorithm based on the social interaction and individual experience. It was proposed by Eberhart and Kennedy in 1995. It has been widely used in finding the solution of optimization problems. This algorithm is inspired by social behaviour of bird flocking or fish schooling. In PSO, the potential solutions, called particles, fly through the search space of the problem by following the current optimum particles. PSO is initialized with a population of random particles positions (solutions) and then searches for optimum in generation to generation. In every iteration, each particle is updated with two best positions (solutions). The first one is the best position (solution) that has been reached so far by the particle and this best position is said to be personal best position and called $p_i^{(k)}$. The other one is the current position (solution), obtained so far by any particle in the population. This best value is a global best and called $p_g^{(k)}$.

In each generation, the velocity and position of ith ($i = 1, 2, \dots .p\_size$) particle are updated by the following rules:

$$v_i^{(k+1)} = w v_i^{(k)} + c_1 r_1 \left(p_i^{(k)} - x_i^{(k)} \right) + c_2 r_2 \left(p_g^{(k)} - x_i^{(k)} \right) \text{ and } x_i^{(k+1)} = x_i^{(k)} + v_i^{(k+1)}$$

Where w is the inertia weight; $k (= 1, 2 \dots m - gen)$ indicates the iterations (generations). The constant $c_1 (> 0)$ and $c_2 (> 0)$ are cognitive learning and social learning rates, respectively, which are the acceleration constants responsible for varying the particle velocity towards $p_i^{(k)}$ and $p_g^{(k)}$ respectively.

Updated velocity of ith particle is calculated by considering three components: (i) previous velocity of the particle, (ii) the distance between the particles best previous and current positions, and (iii) the distance between swarms best experience (the position of the best particle in the swarm) and the current position of the particle.

The velocity is also limited by the range $\left[-v_{max}, v_{max} \right]$ where v_{max} is called the maximum velocity of the particle. The choice of a too small value for v_{max} can cause very small updating of velocities and positions of particles at each iteration. Hence, the algorithm may take a long time to converge and faces the problem of getting stuck to local minima. To overcome this issue, Clerc (1999), Clerc and Kennedy (2002) proposed an improved velocity update rule employing a constriction factor of χ. According to them, the updated velocity is given by

$$v_i^{(k+1)} = \chi \left[v_i^{(k)} + c_1 r_1 \left(p_i^{(k)} - x_i^{(k)} \right) + c_2 r_2 \left(p_g^{(k)} - x_i^{(k)} \right) \right]$$

Here, the constriction factor χ is expressed as

$$\chi \frac{2}{\left| 2 - \phi - \sqrt{\phi^2 - 4\phi} \right|}, \text{ where } \phi = c_1 + c_2, \phi > 4,$$

Algorithm:

> Step 0: Initialize the PSO parameters: bounds of the decision variables; population with random positions and velocities.
>
> Step 1: Evaluate the fitness of all particles.
>
> Step 2: Keep track of the locations where each individual has its highest fitness so far.
>
> Step 3: Keep track of the position with the global best fitness.
>
> Step 4: Update the velocity of each particle.
>
> Step 5: Update the position of each particle.
>
> Step 6: If the stopping criterion is satisfied, go to Step 7, otherwise go to Step 2.
>
> Step 7: Print the position and fitness of global best particle.
>
> Step 8: End

3.3.3 HILL CLIMBING ALGORITHM

As the name suggests, it is mimicking the process of climbing hill from random point and continue the process if peak not achieved. It is a very basic algorithm that can be used to obtain maximum of a given function by just initializing with a random point and then moving to next point. If functional value of new point is better than the previous it is accepted and the process is continued. But if at some point functional value is smaller at the new point then terminate the process and declare the last point as optimum. But as shown in Figure 3.1, user can get stuck with local maxima if initial point is at the left of the local maxima. Other than this in case of plateau (Figure 3.2) and ridges (Figure 3.3) too this method fails to give global optimum.

3.3.4 SIMULATED ANNEALING

Simulated means mimicking and annealing means process of providing heat to metals and then allow them to settle down. So basically this particular algorithm is inspired by annealing process of metallurgy. It is a probabilistic technique for approximating

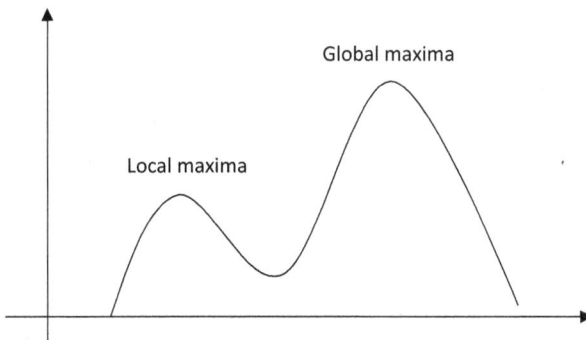

FIGURE 3.1   Graph of an arbitrary function with local and global maxima.

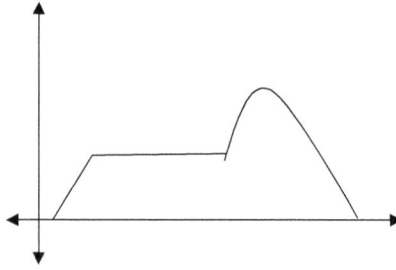

FIGURE 3.2   Plateau.

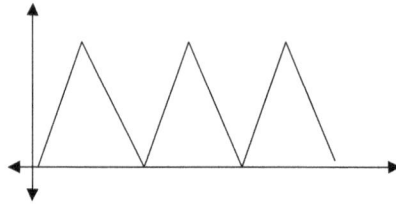

FIGURE 3.3   Ridges.

the global optimum of a given function. It is the extension or a modification of hill climbing where it does not stuck with local optimum as it has provision of downward hill movement with upward. It is a meta-heuristic approach as approximate global optimum in a large search space. For problems where finding an approximate global optimum is more important than finding a precise local optimum in a fixed amount of time, simulated annealing may be preferable to alternatives such as gradient descent. This method can be used for scheduling travelling salesman problem, design of three-dimensional structures of protein molecule in the biotechnology, and printed circuit boards for planning paths for robots, etc.

This process is analogous to the physical annealing of metals. Energy states of metal are cost function in simulated annealing. Similarly, temperature are control parameters and final cooled down crystalline structure is optimal solution. Metal itself is the optimization problem. Here, global optimum can be achieved if cooling process is slow and steady. In other words, moves could be random in the beginning but need to be more precise towards the end to obtain the global optimum.

Working rules for SA are to set initial solution and initial temperature. Then generate new solutions and update the solution. In case it is not acceptable change the temperature. We will continue the process till desired optimal is achieved.

Algorithm

Step 0: Set $x^{(0)}$ and ε as an initial solution and termination criterion, respectively. Fix sufficiently high temperature "T" and number of iterations as "n".

Step1: Calculate a neighbouring point $x^{(t+1)} = Nx^{(t)}$ randomly.

Step 2: If $\Delta E = E(x^{(t+1)}) - E(x^{(t)}) < 0$ for $t = t+1$

Where, ΔE is difference in the energy at $x^{(t+1)}$ and $x^{(t)}$, which is analogous to difference in functional value at two consecutive points.

Else create a random number "r" in the range (0, 1).

If $r \le \exp(-\Delta E / T)$, else go to step 1.

Step 3: If $\left| x^{(t+1)} - x^{(t)} \right| < \varepsilon$ and T is small enough then terminate the process, else if $(t \mod n) = 0$, then accordingly go to step 1.

3.3.5 ANT COLONY OPTIMIZATION ALGORITHM

This algorithm is proposed in 1992 by Macro Dorigo and initially called Ant systems. Since then, many variants of the principle have been developed. It is inspired by the shortest path or trail that ants use to carry their food to home. The main inspiration behind this algorithm is stigmergy comprise of interaction, coordination and adaptation with nature by modifying the local environment. Ant Colony optimization (ACO) takes inspiration from the foraging behaviour of some ant species. These ants deposit pheromone on the ground in order to mark some favourable path that should be followed by other members of the ant colony. Pheromone is a chemical substance released in the environment by animals, specially mammal or insects, that affect the behaviour of other organisms of the same species.

Basically, ants need to find a shortest possible path from source of food to their home. As discussed above this shortest path is achieved through pheromone trails. In the beginning each ant moves in random motion and releases pheromone. This pheromone gets deposited in all the random paths. But the path with more pheromone has the highest probability of getting followed by others. Slowly with time they concentrate on a unique shortest path.

Well known, travelling salesman problem can be solved by this algorithm. In this problem a set of cities is given and distance between each of them is also provided. Aim of the salesman is to find the shortest possible route with the requirement of visiting each city exactly once. Other than travelling salesman problem it can also be useful in quadratic assignment problems, network model problem, vehicle routing, and graph colouring.

Algorithm

Step 0: Explore all possible paths.
Step 1: Ant "k" at node "h" chooses node "s" to move using:

$$s \begin{cases} \left\{ \max_{w \notin M_k} \left\{ [\tau(h,u)] \bullet [\eta(h,u)]^{\beta} \right\} \quad if \left\{ q \le q_0 \right\} \right. \\ \qquad S \qquad\qquad\qquad otherwise \end{cases}$$

Here S is random variable which favours shorter edges with higher level of pheromone trail through a probability distribution mentioned below.

$$P_k(h,s) = \begin{cases} \dfrac{[\tau(h,u)] \bullet [\eta(h,u)]^\beta}{\displaystyle\sum_{w \notin M_k} [\tau(h,u)] \bullet [\eta(h,u)]^\beta} , if\left\{ s \notin M_k \right\} \\ \\ 0 \qquad\qquad\qquad otherwise \end{cases}$$

Note that $\left[\tau(h,u)\right]$ is amount of pheromone trail on edge (h,u) whereas $\eta(h,u)$ is heuristic function on edge (h,u).

Step 2: Pheromone amount on the edges are updated locally as well as globally.

Step 3: Global updating rewards edged belongs to shortest tours. Once ants have completed their routes, ant that has travelled shortest path deposits additional pheromone on each visited edge.

$$\phi(h,s) \leftarrow (1-\alpha) \bullet \phi(h,s) + \alpha \bullet \Delta\phi(h,s)$$

Step 4: Every time a path is chosen, pheromone gets updated.

$$\tau(h,s) \leftarrow (1-\alpha) \bullet \tau(h,s) + \alpha \bullet \tau_o.$$

Advantages:

1. Perform better than other global optimization techniques such as genetic algorithm, neural network, and simulated annealing for a particular class of problems.
2. Many modified versions are available and can be used in dynamic applications.
3. Convergence is sure and certain.

Disadvantages:

1. For small set of nodes, many other search techniques are available with less computational cost.
2. Although convergence is certain but time is uncertain.
3. Some of the time computational cost is high.

3.3.6 TABU SEARCH ALGORITHM

It is a meta-heuristic approach that guides local heuristic search procedure to explore the solution space for global optimum by using a Tabu list. It is dynamic in nature and uses flexible memory to restrict the solution choice to some subset neighbourhood of current solution by strategic restrictions. At the same time it explores search space with appropriate aspiration level. The main features of Tabu search is control and have a check on what enters in the Tabu list called as Forbidding Feature. It controls exits from Tabu list known as Freeing Feature and most important one is a proper coordination between the forbidding and freeing strategy to select the trial solutions which is better known as short-term strategy. To identify neighbouring or adjacent solutions a "neighbourhood" is constructed to move to another solutions from the current solution. Choosing a particular solution from the neighbourhood depends on search history and on frequency of solution called attributes that have already produced past solutions. As mentioned earlier this algorithm has a flexible memory and therefore it records forbidden moves for future known as tabu moves. There is provision of exceptions too in aspiration criterion. When a tabu move gives a better result compared to all the solutions received so far, then it can be overridden.

Stopping Criterion

1. There is no feasible solution in the neighbourhood of solution k of the present iteration.
2. Maximum number of iterations are achieved already.
3. Improvement of the solution is not significant or less than a prior defined (very small) number.

Algorithm

Step 0: Choose an initial solution k. Set $k^* = k, m = 0$

Step 1: Take $k = k+1$ and produce $V^* \ni V^* \subset N(k,m)$ with the condition that either one of the tabu condition gets violated or one of the aspiration condition holds.

Step 2: Choose a best possible n in V^* and set $m = n$.

Step 3: If $f(k^*) < f(k)$ then set $k^* = k$.

Step 4: Update Tabu and aspiration conditions accordingly.

Step 5: If stopping criterion is achieved terminate the process or else go to step 1.

Advantages:

1. Rejects non-improving solutions to avoid local minimum instead of global.
2. Useful in case of both continuous and discrete solution spaces.
3. Very useful for complex problems of scheduling, vehicle routing, and quadratic assignment where other approaches either fail or stuck with local optimum.

Disadvantages:

1. Large number of parameters are required to be determined.
2. Computational cost is high.
3. Number of iterations can be large.
4. Good parameter setting is required to achieve the global optimum.

TRY YOURSELF

Q1. Obtain the $\underset{x}{Min}\, f = (x_1 - 1)^2 + (x_2 - 5)^2$

Subject to $\quad \begin{aligned} g_1 &= -x_1^2 + x_2 - 4 \leq 0 \\ g_2 &= -(x_1 - 2)^2 + x_2 - 3 \leq 0 \end{aligned} \quad$ using KT conditions.

Q2. Write code to solve constrained problem using Genetic Algorithm in C program.

Q3. Compare advantages and disadvantages of Hill Climbing and Simulated Annealing. Do they have any relation with each other?

Q4. Ant colony algorithm is most efficient with which kind of problems. Explain with suitable illustration.

Q5. Find the minimum of $f = x^5 - 5x^3 - 20x + 5$ in the range (0, 3) using the ant colony optimization method. Show detailed calculations for two iterations with four ants.

4 Applications of Non-Linear Programming

4.1 BASICS OF FORMULATION

In the previous three chapters, we discussed various conventional as well as modern approaches for optimizing a given non-linear programming problem. In Section 4.3 of this chapter we will also see various inbuilt functions that can be used to solve a large set of the problem. But actual challenge lies in mathematical formulation of real world problem. Non-linear problem exists in management, core sciences, engineering, medical, military, and finance. It could be a problem for cost minimization or profit maximization, or designing that lead to profit maximization.

Formulation of any real world problem needs sound knowledge of that field as well as NLP. Both together help in establishing relationship between various design (decision variables) and thus concluding with correct objective function. Establishing authentic constraint and bounds is equally important. Here, in the next section we have demonstrated some illustrations to show formulation of NLP in various fields.

4.2 EXAMPLES OF NLP FORMULATION

EXAMPLE 1: PROFIT MAXIMIZATION – PRODUCTION PROBLEM

A manufacturer of coloured televisions is planning the introduction of two new products: a 19-inch stereo colour set with a manufacturer's suggested retail price of $339 per year, and a 21-inch stereo colour set with a suggested retail price of $399 per year. The cost of the company is $195 per 19-inch set and $225 per 21-inch set, plus additional fixed costs of $400,000 per year. In the competitive market, the number of sales will affect the sales price. It is estimated that for each type of set, the sales price drops by one cent for each additional unit sold. Furthermore, sales of the 19-inch set will affect sales of the 21-inch set and vice-versa. It is estimated that the price of 19-inch set will be reduced by an additional 0.3 cents for each 21-inch set sold, and the price of 21-inch set will decrease by 0.4 cents for each 19-inch set sold. The company believes that when the number of units of each type produced is consistent with these assumptions all units will be sold. How many units of each type of set should be manufactured such as the profit of company is maximized?

Solution:

The relevant variables of this problem are:

s_1: Number of units of the 19-inch set produced per year
s_2: Number of units of the 21-inch set produced per year
p_1: Sales price per unit of the 19-inch set ($)
p_2: Sales price per unit of the 21-inch set ($)
C: Manufacturing costs ($ per year),
R: Revenue from sales ($ per year),
P: Profit from sales ($ per year).

The market estimates results in the following model equations,

$$p_1 = 339 - 0.01s_1 - 0.003s_2$$
$$p_2 = 399 - 0.04s_1 - 0.001s_2$$
$$R = s_1 p_1 + s_2 p_2$$
$$C = 400,000 + 195s_1 + 225s_2$$
$$P = R - C$$

The profit then becomes a non-linear function of (s_1, s_2),

$$P(s_1, s_2) = -400,000 + 144s_1 + 174s_2 - 0.01s_1^2 - 0.07s_1 s_2 - 0.01s_2^2$$

If the company has unlimited resources, the only constraints are $s_1, s_2 \geq 0$.

Unconstrained Optimization. We first solve the unconstrained optimization problem. If P has a maximum in the first quadrant this yields the optimal solution. The condition for an extreme point of P leads to a linear system of equations for (s_1, s_2),

$$\frac{\partial P}{\partial s_1} = 144 - 0.02s_1 - 0.007s_2 = 0$$

$$\frac{\partial P}{\partial s_2} = 174 - 0.007s_1 - 0.02s_2 = 0$$

The solution of these equations is $s_1^* = 4735$, $s_2^* = 7043$ with profit value $P^* = P(s_1^*, s_2^*) = 553,641$. Since s_1^*, s_2^* are positive, the inequality constraints are satisfied. To determine the type of the extreme point, we inspect the Hessian matrix,

$$HP(s_1^*, s_2^*) = \begin{bmatrix} -0.02 & -0.007 \\ -0.007 & -0.02 \end{bmatrix}$$

A sufficient condition for a maximum is that $(HP)_{11} < 0$ and $\det(HP) > 0$. Both of these conditions are satisfied and so our solution point is indeed a maximum, in fact a global maximum.

Constrained Optimization. Now suppose the company has limited resources which restrict the number of units of each type produced per year to

$$s_1 \leq 5,000, \ s_2 \leq 8,000, \ s_1 + s_2 \leq 10,000.$$

The first two constrains are satisfied by (s_1^*, s_2^*), however $s_1^* + s_2^* = 11,278$. The global maximum point of P is now no longer in the feasible region, thus the optimal solution must be on the boundary. We therefore solve the constrained optimization problem
subject to

$$c \ (s_1, s_2) = s_1 + s_2 - 10,000 = 0.$$

We can either substitute s_1 or s_2 from the constraint equation into P and solve an unconstrained one-variable optimization problem, or use Lagrangian multipliers. Choosing the second approach, the equation $\nabla P = \lambda \nabla c$ becomes

$$144 - 0.02s_1 - 0.007s_2 = \lambda$$
$$174 - 0.007s_1 - 0.02s_2 = \lambda,$$

which reduces to a single equation for s_1, s_2. Together with the constraint equation we then have again a system of two linear equations,

$$-0.013s_1 + 0.013s_2 = 30$$
$$s_1 + s_2 = 10,000.$$

The solution is $s_1^* = 3846$, $s_2^* = 6154$ with profit value $P^* = 532,308$.

EXAMPLE 2: COST MINIMIZATION – OPTIMUM DESIGNING PROBLEM

We need to design a "CAN" in such a way that its manufacturing cost can be minimized, assuming manufacturing cost is Rs 9/cm². "CAN" should be designed to hold almost 200 ml (200 cc) of liquid. Assuming r and h as radius and height of the "CAN", $3.3 < r < 5$ and $4.7 < h < 20.5$ (cm) to cater aesthetics and confront of the user $h \geq 3.5r$.

Solution:
Here, our objective is to minimize the total cost of manufacturing "CAN". Cost of fabrication depends on total surface area of "CAN" (cylinder).
∴ Total surface area of cylinder $= 2\pi rh + 2\pi r^2 = 2\pi r(h + r)$ from Figure 4.1

$$\therefore \text{ Total cost} = 2 \ \pi r(h + r) \times 9$$

$$\therefore \text{ Min } f(h, r) = 18 \ \pi r(h + r) \qquad \text{[Objective function]}$$

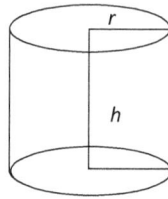

FIGURE 4.1   CAN with radius r and height h.

It is mentioned in the problem that the capacity of "CAN" must be 200 ml (cc)

$$\therefore \text{ Volume of cylinder} = \pi r^2 h = 200 \quad \text{[Equality constraint]}$$

For aesthetic and confrontness of users $h \geq 3.5r$

$$\therefore 3.5r - h \leq 0 \qquad \text{[Inequality constraint]}$$

Other constraints on decision variables are

$$3.3 < r < 5$$

$$4.7 < h < 20.5 \qquad \text{[Boundary conditions]}$$

Finally, mathematical formulation is given as

$$\text{Min } f(h,r) = 18\pi r(h+r)$$

$$\pi r^2 h = 200$$

$$3.5r - h \leq 0$$

$$3.3 < r < 5$$

$$4.7 < h < 20.5$$

where r and h are design (decision) variables

EXAMPLE 3: COST MINIMIZATION – ELECTRICAL ENGINEERING

Optimal power flow problem is a typical example of optimization from the domain of electrical engineering. Cost of generating electricity would be different for various generators depending on the fuel involved, plant efficiency, technology, etc. The objective of the optimal power flow problem is to minimize the total cost of generating electricity for different loads subject to the load demand balance constraints, generation limits, transmission flow limits, and other such system constraints.

To understand this problem, consider the three buses network shown in Figure 4.1. For an optimal power flow problem, the objective function is to minimize the total cost of generation. This is represented in equation (4.1). Individual cost of production for the given system is as follows: $F(P_1) = 0.01P_1^2 + 7.5P_1 + 450$; $F(P_1) = 0.02P_2^2 + 10.5P_1 + 150$; $F(P_1) = 0.15P_1^2 + 4.5P_1 + 600$. These generators have a minimum and maximum generation capacity of 50 MW and 200 MW for generator 1 respectively, 10 MW and 150 MW for generator 2, and 60 MW and 300 MW for generator 3. Line resistance and reactance of all three lines is 0.05 and 0.14 respectively and the load demand is of 400 MW.

Mathematical formulation of this problem is as shown below

$$\min \quad \sum_{g=1}^{3} F\left(P_g\right) \tag{i}$$

$$\text{s.t.} \quad V_1 V_2 \left(\frac{\cos\theta_{12}}{0.05} + \frac{\sin\theta_{12}}{0.14} \right) + V_1 V_3 \left(\frac{\cos\theta_{13}}{0.05} + \frac{\sin\theta_{13}}{0.14} \right) + P_1 = 0 \tag{ii}$$

$$V_2 V_1 \left(\frac{\cos\theta_{12}}{0.05} + \frac{\sin\theta_{12}}{0.14} \right) + V_2 V_3 \left(\frac{\cos\theta_{23}}{0.05} + \frac{\sin\theta_{23}}{0.14} \right) + P_2 = 0 \tag{iii}$$

$$V_3 V_1 \left(\frac{\cos\theta_{13}}{0.05} + \frac{\sin\theta_{13}}{0.14} \right) + V_3 V_2 \left(\frac{\cos\theta_{23}}{0.05} + \frac{\sin\theta_{23}}{0.14} \right) + P_3 = 400 \tag{iv}$$

$$50 \leq P_1 \leq 200 \tag{v}$$

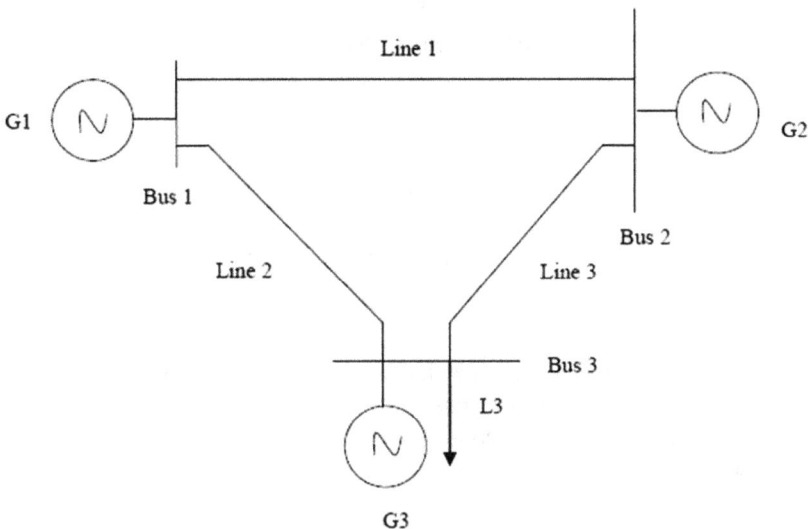

FIGURE 4.2 Simple three bus network.

$$10 \leq P_2 \leq 150 \tag{vi}$$

$$60 \leq P_3 \leq 300 \tag{vii}$$

The goal of this optimization problem is to reduce the total generation cost to a minimum subject to the constraints. There are nine decision variables. Quadratic expression related to variables, P_1, P_2, and P_3 represent the total generation cost. The other variables, voltage terms, V_1, V_2, and V_3 and angles θ_{12}, θ_{23}, and θ_{31} have zero coefficients. For simplicity only three equality constraints (ii), (iii), and (iv) are load balance constraints at each bus. The non-linear terms represent the power flowing through individual transmission line. The last three set of inequality constraints are generation limit constraints. This entire formulation is of the non-linear convex type but it can also be approximated as linear programming model type.

EXAMPLE 4: DESIGN OF A SMALL HEAT EXCHANGER NETWORK – CHEMICAL ENGINEERING

Consider the optimization of the small process network shown in Figure 4.1 with two process streams and three heat exchangers. Using temperatures defined by $T_{in} > T_{out}$ and $t_{out} > t_{in}$, the "hot" stream with a fixed flow rate F and heat capacity C_p needs to be cooled from T_{in} to T_{out}, while the "cold" stream with fixed flow rate f and heat capacity c_p needs to be heated from t_{in} to t_{out}. This is accomplished by two heat exchangers; the heater uses steam at temperature T_s and has a heat duty Q_h, while the cooler uses cold water at temperature T_w and has a heat duty Q_c. However, considerable energy can be saved by exchanging heat between the hot and cold streams through the third heat exchanger with heat duty Q_m and hot and cold exit temperatures, T_m and t_m, respectively. The model for this system is given as follows:

- The energy balance for this system is given by

$$Q_c = FC_p \left(T_m - T_{out} \right), \tag{4.1}$$

$$Q_h = fc_p \left(t_{out} - t_m \right), \tag{4.2}$$

$$Q_m = fc_p \left(t_m - t_{in} \right) = FC_p \left(T_{in} - T_m \right). \tag{4.3}$$

- Each heat exchanger also has a capital cost that is based on its area $A_i, i \in \{c, h, m\}$, for heat exchange. Here we consider a simple countercurrent, shell, and tube heat exchanger with an overall heat transfer coefficient, $U_i, i \in \{c, h, m\}$. The resulting area equations are given by

$$Q_i = U_i A_i \, \Delta T_{lm}^i, i \in \{c, h, m\}. \tag{4.4}$$

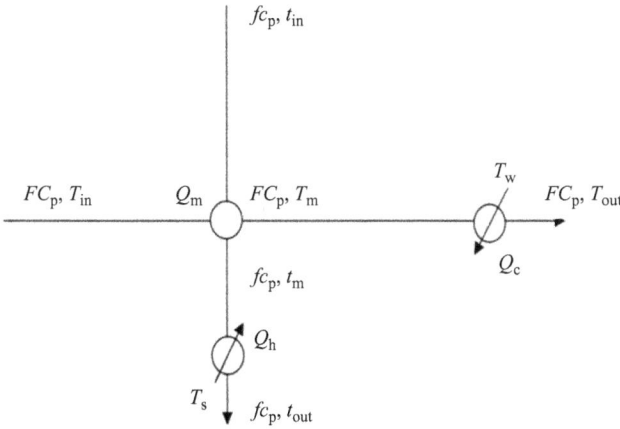

FIGURE 4.3 Example of simple heat exchanger network.

- The log-mean temperature difference ΔT_{lm}^i is given by

$$\Delta T_{lm}^i = \frac{\Delta T_a^i - \Delta T_b^i}{\ln\left(\Delta T_a^i / \Delta T_b^i\right)}, i \in \{c,h,m\}, \tag{4.5}$$

and, $\Delta T_a^c = T_m - T_w, \qquad \Delta T_b^c = T_{out} - T_w,$

$$\Delta T_a^h = T_s - t_m, \qquad \Delta T_b^h = T_s - t_{out},$$

$$\Delta T_a^m = T_{in} - t_m, \quad \Delta T_b^m = T_m - t_{in}.$$

Our objective is to minimize the total cost of the system, i.e., the energy cost as well as the capital cost of the heat exchangers. This leads to the following NLP:

$$\sum_{i \in \{c,h,m\}} \left(\hat{c}_i Q_i + \bar{c}_i A_I^\beta \right) \tag{4.6}$$

$$\text{s.t } (4.1)\text{-}(4.5). \tag{4.7}$$

$$Q_i \geq 0, \Delta T_a^i \geq \Delta, \Delta T_b^i \geq \Delta \tag{4.8}$$

where the cost coefficients $\hat{c}_i$ and $\bar{c}_i$ reflect the energy and amortized capital prices, the exponent $\beta \in (0,1]$ reflects the economy of scale of the equipment, and a small constant $\Delta > 0$ is selected to prevent the log-mean temperature difference from becoming undefined. This example has one degree of freedom. For instance, if the heat duty Q_m is specified, then the hot and cold stream temperatures and all of the remaining quantities can be calculated.

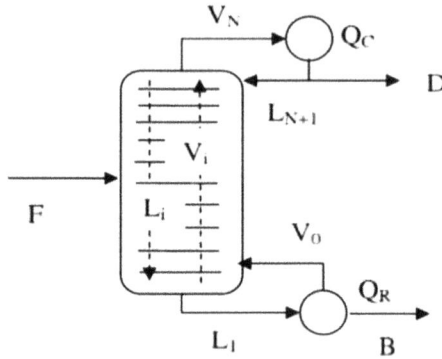

FIGURE 4.4 Distillation column example.

EXAMPLE 5: REAL-TIME OPTIMIZATION OF A DISTILLATION COLUMN – PETROLEUM ENGINEERING

Distillation is the most common means for separation of chemical components and lies at the heart of petroleum refining process; it has no moving parts and scales easily and economically to all production levels. However, distillation is highly energy intensive and can consume 80%–90% of the total energy in a typical chemical or petrochemical process. As a result, optimization of distillation columns is essential for the profitability of these processes. Moreover, because distillation feeds, product demands, and even ambient conditions change overtime, the real-time optimization in response to these changes is also a key contributor to successful operation. Consider the distillation column shown in Figure 4.2 with N trays. As seen in the figure, liquid and vapour contact each other and approach equilibrium (i.e., boiling) on each tray. Moreover, the counter current flow of liquid and vapour provides an enrichment of the volatile (light) components in the top product and the remaining components in the bottom product. Two heat exchangers, the top condenser and the bottom reboiler, act as sources for the condensed liquid vapour and boiled-up vapour, respectively. The hydrocarbon feed contains chemical components given by the set

$$C = \left\{ propane, isobutane, n-butane, isopentane, n-pentane \right\}.$$

The column is specified to recover most of then-butane (the light key) in the top product and most of the isopentane (the heavy key) in the bottom product. We assume a total condenser and partial reboiler, and that the liquid and vapour phases are in equilibrium. A tray-by-tray distillation column model is constructed as follows using the MESH (Mass–Equilibrium–Summation–Heat) equations:

Total Mass Balances

$$B + V_0 - L_1 = 0, \tag{4.9}$$

$$L_i + V_i - L_{i+1} - V_{i-1} = 0, \quad i \epsilon [1,N], i \notin S, \tag{4.10}$$

$$L_i + V_i - L_{i+1} - V_{i-1} - F = 0, \quad i \epsilon \, S, \tag{4.11}$$

$$L_{N+1} + D - V_N = 0. \tag{4.12}$$

Component Mass Balances

$$Bx_{0,j} + V_0 y_{0,j} - L_1 x_{1,j} = 0, \quad j \epsilon \, C, \tag{4.13}$$

$$L_i x_{i,j} - V_i y_{i,j} - L_{i+1} x_{i+1,j} - V_{i-1} y_{i-1,j} = 0, \, j \epsilon \, C, i \epsilon \left[1, N\right], i \notin S, \tag{4.14}$$

$$L_i x_{i,j} - V_i y_{i,j} - L_{i+1} x_{i+1,j} - V_{i-1} y_{i-1,j} - Fx_{F,j} = 0, \, j \epsilon \, C, i \epsilon \, S, \tag{4.15}$$

$$\left(L_{N+1} + D\right) x_{N+1,j} - V_N y_{N,j} = 0, j \epsilon \, C, \tag{4.16}$$

$$x_{N+1,j} - y_{N,j} = 0, j \epsilon \, C, \tag{4.17}$$

Enthalpy Balances

$$BH_B + V_0 H_{V,0} - L_1 H_{L,1} - Q_R = 0 \tag{4.18}$$

$$L_i H_{L,i} - V_i H_{V,i} - L_{i+1} H_{L,i+1} - V_{i-1} H_{V,i-1} = 0, i \epsilon \left[1, N\right], i \notin S, \tag{4.19}$$

$$L_i H_{L,i} - V_i H_{V,i} - L_{i+1} H_{L,i+1} - V_{i-1} H_{V,i-1} - FH_F = 0, i \epsilon \, S, \tag{4.20}$$

$$V_N H_{V,N} - \left(L_{N+1} - D\right) H_{L,D} - Q_C = 0. \tag{4.21}$$

Summation, Enthalpy, and Equilibrium Relations

$$\sum_{j=1}^{m} y_{i,j} - \sum_{j=1}^{m} x_{i,j} = 0, \, i = 0, \dots, N+1 \tag{4.22}$$

$$y_{i,j} - K_{i,j}\left(T_i, P, x_i\right) x_{i,j} = 0, j \epsilon \, C, \quad i = 0, \dots, N+1 \tag{4.23}$$

$$H_{L,i} = \varphi_L\left(x_i, T_i\right), H_{V,i} = \varphi_V\left(y_i, T_i\right), i = 1, \dots, N \tag{4.24}$$

$$H_B = \varphi_L\left(x_0, T_0\right), H_F = \varphi_L\left(x_F, T_F\right), H_{N+1} = \varphi_L\left(x_{N+1}, T_{N+1}\right), \tag{4.25}$$

where

i	tray index numbered starting from reboiler ($=1$)
$j \in C$	components in the feed. The most volatile (lightest) is *propane*
P	pressure in the column
$S \in [1, N]$	set of feed tray locations in column, numbered from the bottom
F	feed flow rate
L_i / V_i	flow rate of liquid/vapour leaving tray i
T_i	temperature of tray i
H_F	feed enthalpy
$H_{L,i} / H_{V,i}$	enthalpy of liquid/vapour leaving tray i
x_F	feed composition
$x_{i,j}$	mole fraction j in liquid leaving tray i
$y_{i,j}$	mole fraction j in vapour leaving tray i
$K_{i,j}$	non-linear vapour/liquid equilibrium constant
φ_V / φ_L	non-linear vapour/liquid enthalpy function
D / B	distillate/bottoms flow rate
Q_R / Q_C	heat load on reboiler/condenser
$lk / hk \in C$	light/heavy key components that determine the separation.

The feed is a saturated liquid with component mole fractions specified in the order given above. The column is operated at a constant pressure, and we neglect pressure drop across the column. This problem has 2 degrees of freedom. For instance, if the flow rates for V_0 and L_{N+1} are specified, all of the other quantities can be calculated from equations (4.9)–(4.23). The objective is to minimize the reboiler heat duty which accounts for a major portion of operating costs, and we specify that the mole fraction of the light key must be 100 times smaller in the bottom than in the top product. The optimization problem is therefore given by

$$\text{Min } Q_R$$

$$\text{s.t. } (4.9)\text{-}(4.23)$$

$$x_{bottom,lk} \leq 0.01 x_{top,lk},$$

$$L_i, V_i, T_i \geq 0, \, i = 1, \ldots, N+1,$$

$$D, Q_R, Q_C \geq 0,$$

$$x_{i,j} \, y_{i,j} \in [0,1], \, j \in C, \, i = 1, \ldots, N+1$$

Distillation optimization is an important and challenging industrial problem and it also serves as a very useful testbed for non-linear programming. This challenging application can be scaled up in three ways; through the number of trays to increase overall size, through the number of components to increase the size of the equations per tray, and through phase equilibrium relations, which also increase the non-linearity and number of equations on each tray.

4.3 SOLVING NLP THROUGH MATLAB INBUILT FUNCTIONS

CASE 1: To minimize one-variable function within the given bounds

- *Write a m.file for the given function*
- *Define <mathgraphic id="9780367613280_EQ_0892"/> and <mathgraphic id="9780367613280_EQ_0893"/> and then call fminbnd*
- *Use* Golden Section search and parabolic interpolation
- *[x, fval]= fminbnd(fun, x_1, x_2)*

Example:
Find minimum of $f(x) = x^3 - 2x - 5$ in the interval of (0, 2)

Step 1: Write function file in script and save it appropriately (name of your function is "minimf" in the example, it can be changed as per user wish):

```
function F=minimf(x)
F=x^3-2*x-5;
end
```

Step 2: Call *fminbnd by appropriate syntax to calculate both minima and minimum value.*
Calling syntax and output is shown below.

```
>> [x, fval]=fminbnd(@minimf, 0,2)

x =

    0.8165

fval =

   -6.0887
```

CASE 2: To minimize unconstraint multivariable function
Multivariable problem of $\underset{x}{Min\, f(x)}$ kind can be solved using the following inbuilt functions in MATLAB

1. **fminunc** – *It is suitable for continuous functions that have first- and second-order derivatives. It uses quasi-Newton algorithm.*
   - *Write a function m.file for given function.*
   - *Define x_0 (initial approximation) and then call fminunc in a script file using syntax: [x, fval] = fminunc(@myfun, x_0)*

- Example: $Min\ f(x) = 3x_1^2 + 2x_1x_2 + x_2^2, x_0 = [1,1]^T$
- Function file:

```
function F=minimf(x)
  F=3*x(1)^2+2*x(1)*x(2)+x(2)^2;
  end
```

- OUTPUT

```
>> [x,fval]=fminunc(@minimf,[1 1])

x =

   1.0e-006 *

     0.2541    -0.2029

fval =

   1.3173e-013
```

2. ***fminsearch*** *– It can handle even discontinuous functions. It does not need derivative information.*
   - *Write a m.file for function*
   - *Define x_0 (initial guess)and then call fminsearch in a script file.*
   - *Call the fminsearch by following syntax:*
     $x = fminsearch(@myfun, x_0)$
   - *Min $f(x) = 100(x_2 - x_1^2)^2 + (1 - x_1)^2; X_0 = [-1.2, 1]$*
   - INPUT

```
>> x0=[-1.2 1];
>> [x, fval]=fminsearch(@minimf, x0)

x =

     1.0000     1.0000

fval =

   8.1777e-010
```

OUTPUT

output
$$X = 0.2578 \quad 0.2578$$
$$resnorm = 124.3622$$

3. *lsqnonlin – This minimization inbuilt function is specially programmed for functions like* $\min_x \|f(x)\|^2 = \min_x \left(f_1^2(x) + f_2^2(x) + \ldots f_n^2(x) \right)$ *for non-linear least square curve fitting functions*
   - *Write an m.file for given function*
   - *Call lsqnonlin using syntax: x = lsqnonlin(@fun,x0)*
   - Example: Find x that minimizes $\sum_{k=1}^{10} (2 + 2k - e^{kx_1} - e^{kx_2})^2 \ X_0 = [0.3, 0.4]$
   - Function file:

```
function F = myfun25(x)
   k = 1:10;
   F = 2 + 2*k-exp(k*x(1))-exp(k*x(2));
end
```

   - Calling script:

```
>> [x, resnorm]=lsqnonlin(@myfun25, [0.3, 0.4])
Optimization terminated: norm of the current step is less
 than OPTIONS.TolX.

x =

     0.2578     0.2578

resnorm =

   124.3622
```

   - Output

output

$X = 0.2578 \quad 0.2578$

resnorm $= 124.3622$

CASE 3: To minimize constraint multivariable function

$$\min_x f(x) \text{ such that } \begin{cases} c(x) \le 0 \\ ceq(x) - 0 \\ A \cdot x \le b \\ Aeq \cdot x = beq \\ lb \le x \le ub, \end{cases}$$

   - *Write an m.file for given function*
   - *Call lsqnonlin using syntax: [x,fval] = fmincon(@myfun,x₀,A,b)*

- *Example:*
$$Min\ f(x) = -x_1 x_2 x_3 \quad ; X_0 = [10;10;10]$$
$$s.t.\ 0 \le x_1 + 2x_2 + 2x_3 \le 72$$

- *Function file:*

```
function F = myfun25(x)
F = -x(1)*x(2)*x(3);
end
```

- *Calling file:*

```
>> A=[-1 -2 -2; 1 2 2];
>> b=[0; 72];
>> [x, fval]= fmincon(@myfun25, [10; 10; 10]
Warning: Trust-region-reflective method does
 using active-set (line search) instead.
> In fmincon at 422
Optimization terminated: magnitude of direct
 direction less than 2*options.TolFun and ma
 is less than options.TolCon.
Active inequalities (to within options.TolCo
    lower         upper       ineqlin      ineqnonlin
                                2

x =

   24.0000
   12.0000
   12.0000


fval =

 -3.4560e+003
```

- *Output:*

$$output\ x =$$
$$24.0000$$
$$12.0000$$
$$12.0000$$
$$fval = -3.4560e+03$$

CASE 4: Genetic Algorithm through MATLAB

Many heuristic search algorithms such as genetic algorithm, simulated annealing, and PSO are inbuilt in MATLAB. There is provision of customization of the algorithms as well. Beginners can use them instead of writing complete programs by themselves. Here, an illustration is demonstrated to minimize a function with constraint through GA. MATLAB has optimtool box that can be explored for other inbuilt functions.

$$\text{Example: } f(x) = \begin{cases} -\exp[(-x/20)^2], & x \le 20 \\ -\exp(-1)+(x-20)(x-22), & x>20 \end{cases}$$

- Write function file

```
function f=goodfunc(x)
if x<=20
    f=-exp(-(x/20).^2)
else
    f=-exp(-1)+(x-20)*(x-22)
end
```

- Call GA through interacting window of OPTIMTOOL BOX

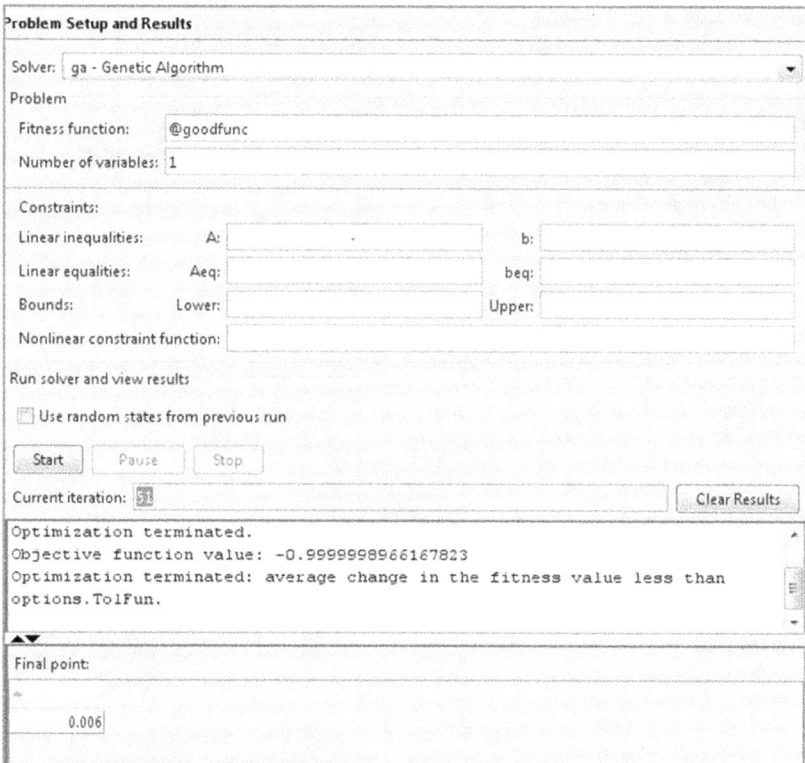

Problem Setup and Results

Solver: ga - Genetic Algorithm

Problem

Fitness function: @goodfunc

Number of variables: 1

Constraints:

Linear inequalities: A: b:

Linear equalities: Aeq: beq:

Bounds: Lower: Upper:

Nonlinear constraint function:

Run solver and view results

☐ Use random states from previous run

Start   Pause   Stop

Current iteration: 51   Clear Results

Optimization terminated.
Objective function value: -0.9999998966167823
Optimization terminated: average change in the fitness value less than options.TolFun.

Final point:

0.006

Here solver used is GA, fitness function is objective function with name "@goodfun". Number of variable is one. Since there is no constraint and no predefined bounds on the decision variable, we can run the solver. It shows that it took 51 iterations with value of variable as 0.006 and functional value at minima is –0.99. Now let us discuss one example with constraints.

Example:

$$Min \ f(x) = 100(x_1^2 - x_2)^2 + (1 - x_2)^2$$
$$subject: x_1 x_2 + x_1 - x_2 + 1.5 \leq 0, \quad 10 - x_1 x_2 \leq 0, \quad 0 \leq x_1 \leq 1, \ 0 \leq x_2 \leq 13$$

Function file:

```
function f=goodfunc1(x)
f=100*(x(1)^2-x(2))^2+(1-x(2))^2;
end
```

Constraint file:

```
function [c, ceq]=constfunc(x)
c=[x(1)*x(2)+x(1)-x(2)+1.5; -x(1)*x(2)+10];
ceq=[];
end
```

OPTIMTOOL Interactive window:

Problem Setup and Results

Solver: ga - Genetic Algorithm

Problem

  Fitness function: @goodfunc1

  Number of variables: 2

  Constraints:

  Linear inequalities: A: b:

  Linear equalities: Aeq: beq:

  Bounds: Lower: [0 0]   Upper: [1 13]

  Nonlinear constraint function: @constfunc

Run solver and view results

  ☐ Use random states from previous run

  [Start]   [Pause]   [Stop]

Current iteration: 4   [Clear Results]

```
Optimization terminated.
Objective function value: 13706.108518470113
Optimization terminated: average change in the fitness value less than
options.TolFun
 and constraint violation is less than options.TolCon.
```

Final point:

1	2
0.812	12.312

Solver is GA, fitness function is objective function, number of variables are 2, bounds are mentioned. Please note that equality constraints can be mentioned directly in interactive window in array form whereas for non-linear constraint we need to write function file and call it appropriately as shown above. Minima is [0.812, 12.312] and minimum is 13706.1085.

4.4   CHOICE OF METHOD

1. Type of problem – linear, non-linear, differentiable or not, constraint, and nature of decision variables.
2. Accuracy of desired solution – local or global optimum is required.
3. Availability of inbuilt programs or need to write customize code pertaining to the problem.
4. Availability of time.

TRY YOURSELF

Q1. Optimize the following function using fsearch:

$$Min\ f(x) = 12(x_2 - x_1)^2 + 5(1 - x_1)^3; X_0 = [0,\ 0]$$

Q2. Optimize the following functions using "fmincon" and "ga" – built function of MATLAB. Initial approximation can be $[1,\ 1]^T$

(i) $Min\ f = (x_1 - 1)^2 + (x_2 - 2)^2 - 4$

Subject to : $x_1 + 2x_2 \leq 5$
$4x_1 + 3x_2 \leq 10$
$6x_1 + x_2 \leq 7, \quad x_1, x_2 \geq 0$

(ii) $Min\ f = 4x_1 + 2x_2 + 3x_3 + 4x_4$

Subject to : $x_1 + x_3 + x_4 \leq 24$
$3x_1 + x_2 + 2x_3 + 4x_4 \leq 48$
$2x_1 + 2x_2 + 3x_3 + 2x_4 \leq 36, \quad x_1, x_2, x_3, x_4 \geq 0$

BIBLIOGRAPHY

1. Rao, S. S., (First published 2009). Engineering and Optimization: Theory and Practice (4th Ed.), New Jersey, U.S.A.: John Wiley & Sons, Inc.
2. Sharma, J. K., (2009). Operations Research: Theory and Practices (4th Ed.), New Delhi, INDIA: Macmilan.
3. Pant, K. K., Sinha, S., & Bajpai, S., (2015). Advances in Petroleum Engineering II – Petrochemical, Studium Press. LLC, U.S.A: Studium Press LLC, U.S.A.
4. Lorenz T. Biegler, (2010). Nonlinear Programming: Concepts, Algorithms and Applications to Chemical Processes, Society for Industrial and Applied Mathematics, U. S.

Index

For Product Safety Concerns and Information please contact our EU
representative GPSR@taylorandfrancis.com
Taylor & Francis Verlag GmbH, Kaufingerstraße 24, 80331 München, Germany